2016 SQA Past Papers & Hodder Gibson Model Papers With Answers

Higher

MATHEMATICS

2014 Specimen Question Paper, Model Paper, 2015 & 2016 Exams

This book contains the official 2014 SQA Specimen Question Paper, 2015 and 2016 Exams for Higher Maths, with associated SQA-approved answers modified from the official marking instructions that accompany the paper.

In addition the book contains a model paper, together with answers, plus study skills advice. This paper, which may include a limited number of previously published SQA questions, has been specially commissioned by Hodder Gibson, and has been written by experienced senior teachers and examiners in line with the new Higher for CfE syllabus and assessment outlines. This is not SQA material but has been devised to provide further practice for Higher examinations.

Hodder Gibson is grateful to the copyright holders, as credited on the final page of the Answer Section, for permission to use their material. Every effort has been made to trace the copyright holders and to obtain their permission for the use of copyright material. Hodder Gibson will be happy to receive information allowing us to rectify any error or omission in future editions.

Hachette UK's policy is to use papers that are natural, renewable and recyclable products and made from wood grown in sustainable forests. The logging and manufacturing processes are expected to conform to the environmental regulations of the country of origin.

Orders: please contact Bookpoint Ltd, 130 Park Drive, Milton Park, Abingdon, Oxon OX14 4SE. Telephone: (44) 01235 827720. Fax: (44) 01235 400454. Lines are open 9.00–5.00, Monday to Saturday, with a 24-hour message answering service. Visit our website at www.hoddereducation.co.uk. Hodder Gibson can be contacted direct on: Tel: 0141 333 4650; Fax: 0141 404 8188; email: hoddergibson@hodder.co.uk

This collection first published in 2016 by
Hodder Gibson, an imprint of Hodder Education,
An Hachette UK Company
211 St Vincent Street
Glasgow G2 5QY

Typeset by Aptara, Inc.

Printed in the UK

A catalogue record for this title is available from the British Library

ISBN: 978-1-4718-9098-7

3 2 1

2017 2016

Introduction

Study Skills – what you need to know to pass exams!

Pause for thought

Many students might skip quickly through a page like this. After all, we all know how to revise. Do you really though?

Think about this:

"IF YOU ALWAYS DO WHAT YOU ALWAYS DO, YOU WILL ALWAYS GET WHAT YOU HAVE ALWAYS GOT."

Do you like the grades you get? Do you want to do better? If you get full marks in your assessment, then that's great! Change nothing! This section is just to help you get that little bit better than you already are.

There are two main parts to the advice on offer here. The first part highlights fairly obvious things but which are also very important. The second part makes suggestions about revision that you might not have thought about but which WILL help you.

Part 1

DOH! It's so obvious but ...

Start revising in good time

Don't leave it until the last minute – this will make you panic.

Make a revision timetable that sets out work time AND play time.

Sleep and eat!

Obvious really, and very helpful. Avoid arguments or stressful things too – even games that wind you up. You need to be fit, awake and focused!

Know your place!

Make sure you know exactly **WHEN and WHERE** your exams are.

Know your enemy!

Make sure you know what to expect in the exam.

How is the paper structured?

How much time is there for each question?

What types of question are involved?

Which topics seem to come up time and time again?

Which topics are your strongest and which are your weakest?

Are all topics compulsory or are there choices?

Learn by DOING!

There is no substitute for past papers and practice papers – they are simply essential! Tackling this collection of papers and answers is exactly the right thing to be doing as your exams approach.

Part 2

People learn in different ways. Some like low light, some bright. Some like early morning, some like evening / night. Some prefer warm, some prefer cold. But everyone uses their BRAIN and the brain works when it is active. Passive learning – sitting gazing at notes – is the most INEFFICIENT way to learn anything. Below you will find tips and ideas for making your revision more effective and maybe even more enjoyable. What follows gets your brain active, and active learning works!

Activity 1 – Stop and review

Step 1

When you have done no more than 5 minutes of revision reading STOP!

Step 2

Write a heading in your own words which sums up the topic you have been revising.

Step 3

Write a summary of what you have revised in no more than two sentences. Don't fool yourself by saying, "I know it, but I cannot put it into words". That just means you don't know it well enough. If you cannot write your summary, revise that section again, knowing that you must write a summary at the end of it. Many of you will have notebooks full of blue/black ink writing. Many of the pages will not be especially attractive or memorable so try to liven them up a bit with colour as you are reviewing and rewriting. **This is a great memory aid, and memory is the most important thing.**

Activity 2 – Use technology!

Why should everything be written down? Have you thought about "mental" maps, diagrams, cartoons and colour to help you learn? And rather than write down notes, why not record your revision material?

What about having a text message revision session with friends? Keep in touch with them to find out how and what they are revising and share ideas and questions.

Why not make a video diary where you tell the camera what you are doing, what you think you have learned and what you still have to do? No one has to see or hear it, but the process of having to organise your thoughts in a formal way to explain something is a very important learning practice.

Be sure to make use of electronic files. You could begin to summarise your class notes. Your typing might be slow, but it will get faster and the typed notes will be easier to read than the scribbles in your class notes. Try to add different fonts and colours to make your work stand out. You can easily Google relevant pictures, cartoons and diagrams which you can copy and paste to make your work more attractive and **MEMORABLE**.

Activity 3 – This is it. Do this and you will know lots!

Step 1

In this task you must be very honest with yourself! Find the SQA syllabus for your subject (www.sqa.org.uk). Look at how it is broken down into main topics called MANDATORY knowledge. That means stuff you MUST know.

Step 2

BEFORE you do ANY revision on this topic, write a list of everything that you already know about the subject. It might be quite a long list but you only need to write it once. It shows you all the information that is already in your long-term memory so you know what parts you do not need to revise!

Step 3

Pick a chapter or section from your book or revision notes. Choose a fairly large section or a whole chapter to get the most out of this activity.

With a buddy, use Skype, Facetime, Twitter or any other communication you have, to play the game "If this is the answer, what is the question?". For example, if you are revising Geography and the answer you provide is "meander", your buddy would have to make up a question like "What is the word that describes a feature of a river where it flows slowly and bends often from side to side?".

Make up 10 "answers" based on the content of the chapter or section you are using. Give this to your buddy to solve while you solve theirs.

Step 4

Construct a wordsearch of at least 10 × 10 squares. You can make it as big as you like but keep it realistic. Work together with a group of friends. Many apps allow you to make wordsearch puzzles online. The words and phrases can go in any direction and phrases can be split. Your puzzle must only contain facts linked to the topic you are revising. Your task is to find 10 bits of information to hide in your puzzle, but you must not repeat information that you used in Step 3. DO NOT show where the words are. Fill up empty squares with random letters. Remember to keep a note of where your answers are hidden but do not show your friends. When you have a complete puzzle, exchange it with a friend to solve each other's puzzle.

Step 5

Now make up 10 questions (not "answers" this time) based on the same chapter used in the previous two tasks. Again, you must find NEW information that you have not yet used. Now it's getting hard to find that new information! Again, give your questions to a friend to answer.

Step 6

As you have been doing the puzzles, your brain has been actively searching for new information. Now write a NEW LIST that contains only the new information you have discovered when doing the puzzles. Your new list is the one to look at repeatedly for short bursts over the next few days. Try to remember more and more of it without looking at it. After a few days, you should be able to add words from your second list to your first list as you increase the information in your long-term memory.

FINALLY! Be inspired...

Make a list of different revision ideas and beside each one write **THINGS I HAVE** tried, **THINGS I WILL** try and **THINGS I MIGHT** try. Don't be scared of trying something new.

And remember – "FAIL TO PREPARE AND PREPARE TO FAIL!"

Higher Mathematics

The course

The Higher Mathematics course aims to:

- motivate and challenge learners by enabling them to select and apply mathematical techniques in a variety of mathematical situations
- develop confidence in the subject and a positive attitude towards further study in mathematics and the use of mathematics in employment
- deliver in-depth study of mathematical concepts and the ways in which mathematics describes our world
- allow learners to interpret, communicate and manage information in mathematical form; skills which are vital to scientific and technological research and development
- deepen learners' skills in using mathematical language and exploring advanced mathematical ideas.

The Higher qualification in Mathematics is designed to build upon and extend learners' mathematical skills, knowledge and understanding in a way that recognises problem solving as an essential skill and enables them to integrate their knowledge of different aspects of the subject.

You will acquire an enhanced awareness of the importance of mathematics to technology and to society in general. Where appropriate, mathematics will be developed in context, and mathematical techniques will be applied in social and vocational contexts related to likely progression routes such as commerce, engineering and science where the mathematics learned will be put to direct use.

The syllabus is designed to build upon your prior learning in the areas of algebra, geometry and trigonometry and to introduce you to elementary calculus.

How the course is assessed

- To gain the Course award, you must pass the three units – Expressions & Functions, Relationships & Calculus, and Applications – as well as the examination.
- The units are assessed internally on a pass/fail basis.
- The examination is set and marked by the SQA.
- The course award is graded A–D, the grade being determined by the total mark you score in the examination.

The examination

- The examination consists of two papers. The number of marks and the times allotted for the papers are:

Paper 1 (non-calculator)	60 marks	1 hour 10 minutes
Paper 2	70 marks	1 hour 30 minutes

- The examination tests skills beyond the minimum competence required for the units. Both papers contain short and extended response questions in which candidates are required to apply numerical, algebraic, geometric, trigonometric, calculus, and reasoning skills.
- The examination is designed so that approximately 65% of the marks will be available for level C responses.
- Some questions will assess only operational skills (65% of the marks) but other questions will require both operational and reasoning skills (35% of the marks).

Further details can be found in the Higher Mathematics section on the SQA website: http://www.sqa.org.uk/sqa/47910.html.

Key tips for your success

Practise! Practise! Practise!

DOING maths questions is the most effective use of your study time. You will benefit much more from spending 30 minutes doing maths questions than spending several hours copying out notes or reading a maths textbook.

Basic skills

You must practise the following essential basic skills for Higher Mathematics throughout the duration of this course – expanding brackets; solving equations; manipulating algebraic expressions; and, in particular, working with exact values with trigonometric expressions and equations.

Non-routine problems

It is important to practise non-routine problems as often as possible throughout the course, particularly if you are aiming for an A or B pass in Higher Mathematics.

Graph sketching

Graph sketching is an important and integral part of Mathematics. Ensure that you practise sketching graphs on plain paper whenever possible throughout this course. Neither squared nor graph paper are allowed in the Higher Mathematics examination.

Marking instructions

Ensure that you look at the detailed marking instructions of past papers. They provide further advice and guidelines as well as showing you precisely where, and for what, marks are awarded.

Show all working clearly

The instructions on the front of the exam paper state that *"Full credit will be given only where the solution contains appropriate working."* A "correct" answer with no working may only be awarded partial marks or even no marks at all. An incomplete answer will be awarded marks for any appropriate working. Attempt every question, even if you are not sure whether you are correct or not. Your solution may contain working which will gain some marks. A blank response is certain to be awarded no marks. Never score out working unless you have something better to replace it with.

Make drawings

Try drawing what you visualise as the "picture", described within the wording of each relevant question. This is a mathematical skill expected of most candidates at Higher level. Making a rough sketch of the diagram in your answer booklet may also help you interpret the question and achieve more marks.

Extended response questions

You should look for connections between parts of questions, particularly where there are three or four sections to a question. These are almost always linked and, in some instances, an earlier result in part (a) or (b) is needed and its use would avoid further repeated work.

Notation

In all questions, make sure that you use the correct notation. In particular, for integration questions, remember to include 'dx' within your integral.

Radians

Remember to work in radians when attempting any calculus questions involving trigonometric functions.

Simplify

Get into the habit of simplifying expressions before doing any further work with them. This should make all subsequent work easier.

Subtraction

Be careful when subtracting one expression from another: ensure that any negative is applied correctly.

Good luck!

Remember that the rewards for passing Higher Mathematics are well worth it! Your pass will help you get the future you want for yourself. In the exam, be confident in your own ability. If you're not sure how to answer a question, trust your instincts and just give it a go anyway – keep calm and don't panic! GOOD LUCK!

HIGHER

2014 Specimen Question Paper

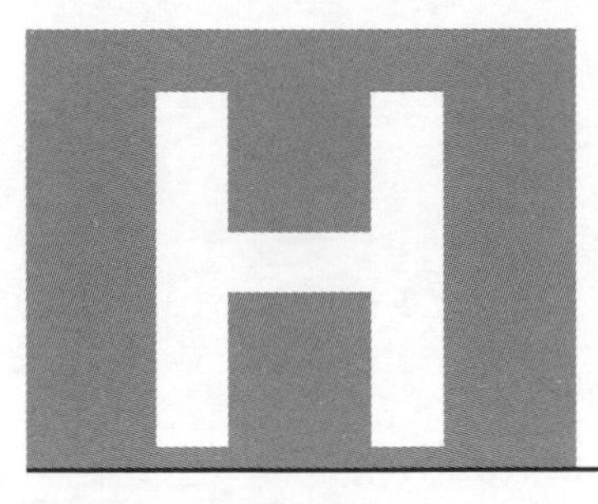

National
Qualifications
SPECIMEN ONLY

SQ30/H/01

Mathematics
Paper 1
(Non-Calculator)

Date — Not applicable

Duration — 1 hour and 10 minutes

Total marks — 60

Attempt ALL questions.

You may NOT use a calculator.

Full credit will be given only to solutions which contain appropriate working.

State the units for your answer where appropriate.

Write your answers clearly in the answer booklet provided. In the answer booklet you must clearly identify the question number you are attempting.

Use **blue** or **black** ink.

Before leaving the examination room you must give your answer booklet to the Invigilator; if you do not you may lose all the marks for this paper.

FORMULAE LIST

Circle:

The equation $x^2 + y^2 + 2gx + 2fy + c = 0$ represents a circle centre $(-g, -f)$ and radius $\sqrt{g^2 + f^2 - c}$.

The equation $(x - a)^2 + (y - b)^2 = r^2$ represents a circle centre (a, b) and radius r.

Scalar Product: $\mathbf{a.b} = |\mathbf{a}||\mathbf{b}| \cos \theta$, where θ is the angle between $\mathbf{a}$ and $\mathbf{b}$

or $\mathbf{a.b} = a_1b_1 + a_2b_2 + a_3b_3$ where $\mathbf{a} = \begin{pmatrix} a_1 \\ a_2 \\ a_3 \end{pmatrix}$ and $\mathbf{b} = \begin{pmatrix} b_1 \\ b_2 \\ b_3 \end{pmatrix}$

Trigonometric formulae:

$$\sin(A \pm B) = \sin A \cos B \pm \cos A \sin B$$
$$\cos(A \pm B) = \cos A \cos B \mp \sin A \sin B$$
$$\sin 2A = 2\sin A \cos A$$
$$\cos 2A = \cos^2 A - \sin^2 A$$
$$= 2\cos^2 A - 1$$
$$= 1 - 2\sin^2 A$$

Table of standard derivatives:

$f(x)$	$f'(x)$
$\sin ax$	$a \cos ax$
$\cos ax$	$-a \sin ax$

Table of standard integrals:

$f(x)$	$\int f(x)dx$
$\sin ax$	$-\frac{1}{a} \cos ax + C$
$\cos ax$	$\frac{1}{a} \sin ax + C$

Attempt ALL questions

Total marks — 60

MARKS

1. Find $\int \frac{3x^3+1}{2x^2}\,dx,\ x \neq 0$. **4**

2. Find the coordinates of the points of intersection of the curve $y = x^3 - 2x^2 + x + 4$ and the line $y = 4x + 4$. **5**

3. In the diagram, P has coordinates $(-6, 3, 9)$,

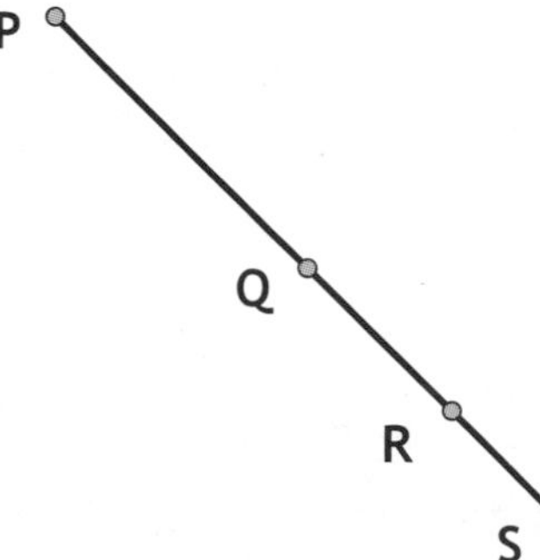

$\overrightarrow{PQ} = 6\mathbf{i} + 12\mathbf{j} - 6\mathbf{k}$ and $\overrightarrow{PQ} = 2\overrightarrow{QR} = 3\overrightarrow{RS}$.

Find the coordinates of S. **5**

4. Given that $2x^2 + px + p + 6 = 0$ has no real roots, find the range of values for p, where $p \in \mathbb{R}$. **4**

5. Line l_1 has equation $\sqrt{3}y - x = 0$.

   (a) Line l_2 is perpendicular to l_1. Find the gradient of l_2. **2**

   (b) Calculate the angle l_2 makes with the positive direction of the x-axis. **2**

6. (a) Find an equivalent expression for $\sin(x + 60)^\circ$. **1**

   (b) Hence, or otherwise, determine the exact value of $\sin 105^\circ$. **3**

7. (a) Show that $(x + 1)$ is a factor of $x^3 - 13x - 12$. **3**

   (b) Factorise $x^3 - 13x - 12$ fully. **2**

MARKS

8. $f(x)$ and $g(x)$ are functions, defined on the set of real numbers, such that

$f(x) = 1 - \frac{1}{2}x$ and $g(x) = 8x^2 - 3$.

(a) Given that $h(x) = g(f(x))$, show that $h(x) = 2x^2 - 8x + 5$. **3**

(b) Express $h(x)$ in the form $a\,(x + p)^2 + q$. **3**

(c) Hence, or otherwise, state the coordinates of the turning point on the graph of $y = h(x)$. **1**

(d) Sketch the graph of $y = h(x) + 3$, showing clearly the coordinates of the turning point and the y-axis intercept. **2**

9. (a) AB is a line parallel to the line with equation $y + 3x = 25$.

A has coordinates $(-1, 10)$.

Find the equation of AB. **1**

(b) $3y = x + 11$ is the perpendicular bisector of AB.

Determine the coordinates of B. **5**

10. Find the rate of change of the function $f(x) = 4\sin^3 x$ when $x = \frac{5\pi}{6}$. **3**

11. The diagram shows the graph of $y = f'(x)$. The x-axis is a tangent to this graph.

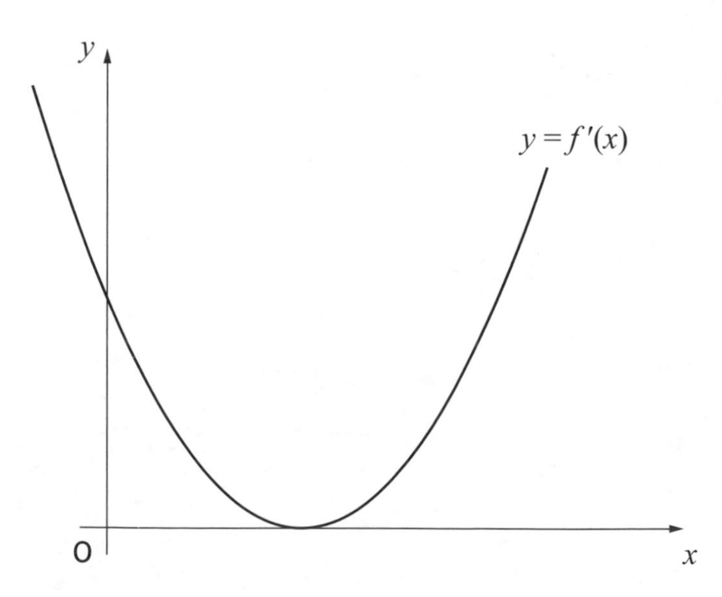

(a) Explain why the function $f(x)$ is never decreasing. **1**

(b) On a graph of $y = f(x)$, the y-coordinate of the stationary point is negative. Sketch a possible graph for $y = f(x)$. **2**

MARKS

12. The voltage, $V(t)$, produced by a generator is described by the function $V(t) = 120\sin 100\pi t$, $t > 0$, where t is the time in seconds.

(a) Determine the period of $V(t)$. **2**

(b) Find the first three times for which $V(t) = -60$. **6**

[END OF SPECIMEN QUESTION PAPER]

[BLANK PAGE]

DO NOT WRITE ON THIS PAGE

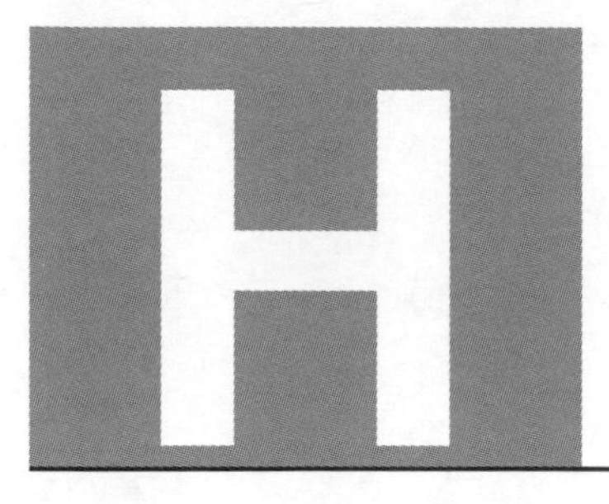

National
Qualifications
SPECIMEN ONLY

SQ30/H/02

Mathematics
Paper 2

Date — Not applicable

Duration — 1 hour and 30 minutes

Total marks — 70

Attempt ALL questions.

You may use a calculator.

Full credit will be given only to solutions which contain appropriate working.

State the units for your answer where appropriate.

Write your answers clearly in the answer booklet provided. In the answer booklet you must clearly identify the question number you are attempting.

Use **blue** or **black** ink.

Before leaving the examination room you must give your answer booklet to the Invigilator; if you do not you may lose all the marks for this paper.

FORMULAE LIST

Circle:

The equation $x^2 + y^2 + 2gx + 2fy + c = 0$ represents a circle centre $(-g, -f)$ and radius $\sqrt{g^2 + f^2 - c}$.

The equation $(x - a)^2 + (y - b)^2 = r^2$ represents a circle centre (a, b) and radius r.

Scalar Product: $\mathbf{a}.\mathbf{b} = |\mathbf{a}||\mathbf{b}| \cos \theta$, where θ is the angle between $\mathbf{a}$ and $\mathbf{b}$

or $\mathbf{a}.\mathbf{b} = a_1b_1 + a_2b_2 + a_3b_3$ where $\mathbf{a} = \begin{pmatrix} a_1 \\ a_2 \\ a_3 \end{pmatrix}$ and $\mathbf{b} = \begin{pmatrix} b_1 \\ b_2 \\ b_3 \end{pmatrix}$

Trigonometric formulae:

$$\begin{aligned}
\sin(\mathrm{A} \pm \mathrm{B}) &= \sin \mathrm{A} \cos \mathrm{B} \pm \cos \mathrm{A} \sin \mathrm{B} \\
\cos(\mathrm{A} \pm \mathrm{B}) &= \cos \mathrm{A} \cos \mathrm{B} \mp \sin \mathrm{A} \sin \mathrm{B} \\
\sin 2\mathrm{A} &= 2\sin \mathrm{A} \cos \mathrm{A} \\
\cos 2\mathrm{A} &= \cos^2 \mathrm{A} - \sin^2 \mathrm{A} \\
&= 2\cos^2 \mathrm{A} - 1 \\
&= 1 - 2\sin^2 \mathrm{A}
\end{aligned}$$

Table of standard derivatives:

$f(x)$	$f'(x)$
$\sin ax$ $\cos ax$	$a \cos ax$ $-a \sin ax$

Table of standard integrals:

$f(x)$	$\int f(x)dx$
$\sin ax$	$-\frac{1}{a} \cos ax + C$
$\cos ax$	$\frac{1}{a} \sin ax + C$

Attempt ALL questions

Total marks — 70

MARKS

1.

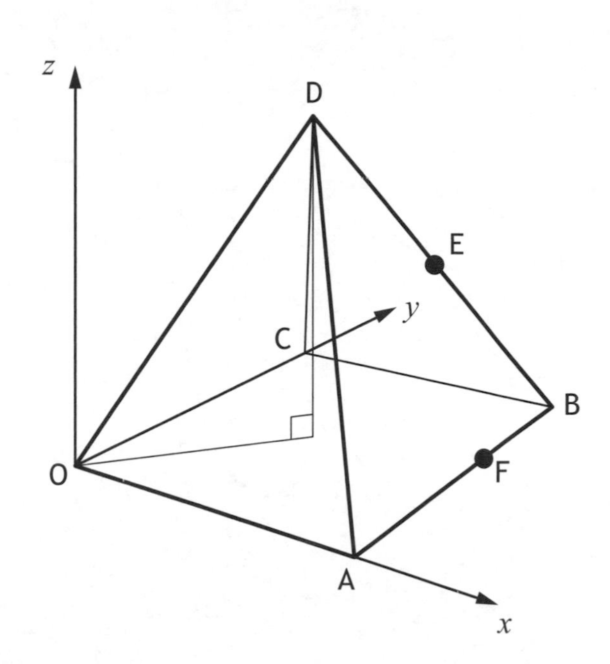

A square based right pyramid is shown in the diagram.

Square OABC has a side length of 60 units with edges OA and OC lying on the x-axis and y-axis respectively.

The coordinates of D are (30, 30, 80).

E is the midpoint of BD and F divides AB in the ratio 2:1.

(a) Find the coordinates of E and F. **2**

(b) Calculate $\overrightarrow{ED}.\overrightarrow{EF}$. **2**

(c) Hence, or otherwise, calculate the size of angle DEF. **4**

2. A wildlife reserve has introduced conservation measures to build up the population of an endangered mammal. Initially the reserve population of the mammal was 2000. By the end of the first year there were 2500 and by the end of the second year there were 2980.

It is believed that the population can be modelled by the recurrence relation:

$u_{n+1} = au_n + b$,

where a and b are constants and n is the number of years since the reserve was set up.

(a) Use the information above to find the values of a and b. **4**

(b) Conservation measures will end if the population stabilises at over 13 000. Will this happen? Justify your answer. **3**

MARKS

3. The diagram shows the graph of $f(x) = x(x - p)(x - q)^2$.

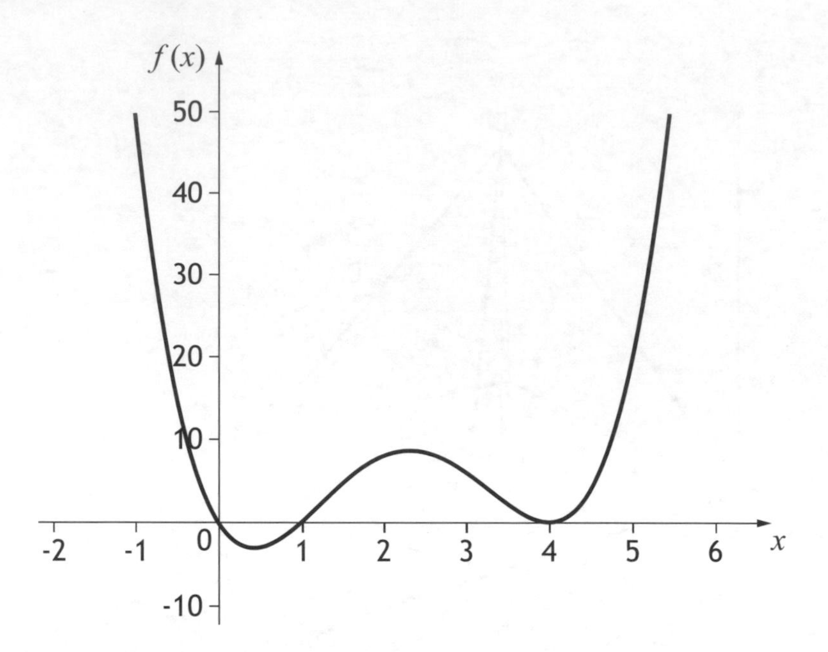

(a) Determine the values of p and q. **1**

(b) Find the equation of the tangent to the curve when $x = 1$. **4**

4. (a) Express $y = \log_4 2x$ in the form $y = \log_4 x + k$, clearly stating the value of k. **2**

(b) Hence, or otherwise, describe the relationship between the graphs of $y = \log_4 2x$ and $y = \log_4 x$. **1**

(c) Determine the coordinates of the point where the graph of $y = \log_4 2x$ intersects the x-axis. **2**

(d) Sketch and annotate the graph of $y = f^{-1}(x)$, where $f(x) = \log_4 2x$. **3**

MARKS

5.

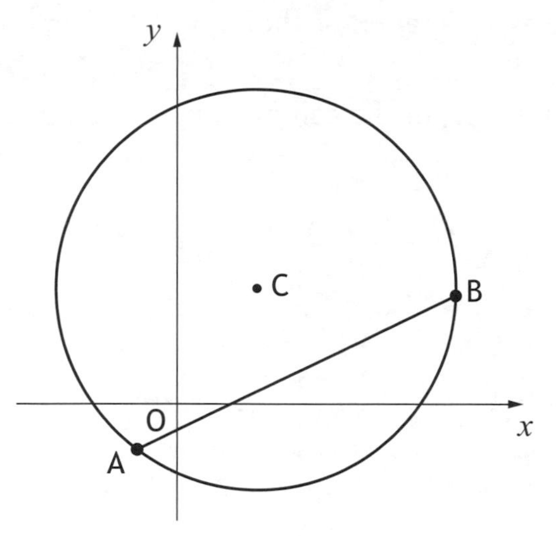

Points A(–1, –1) and B(7, 3) lie on the circumference of a circle with centre C, as shown in the diagram.

(a) Find the equation of the perpendicular bisector of AB. 4

CB is parallel to the x-axis.

(b) Find the equation of the circle, passing through A and B, with centre C. 4

6. The points A(0, 9, 7), B(5, –1, 2), C(4, 1, 3) and D(x, –2, 2) are such that AB is perpendicular to CD.

Determine the value of x. 5

7. Given that $P(t) = 30e^{t-2}$ decide whether each of the statements below is true or false. Justify your answers.

Statement A $P(0) = 30$.

Statement B When $P(t) = 15$, the only possible value of t is 1·3 to one decimal place. 6

MARKS

8. A design for a new grain container is in the shape of a cylinder with a hemispherical roof and a flat circular base. The radius of the cylinder is r metres, and the height is h metres.

The volume of the **cylindrical** part of the container needs to be 100 cubic metres.

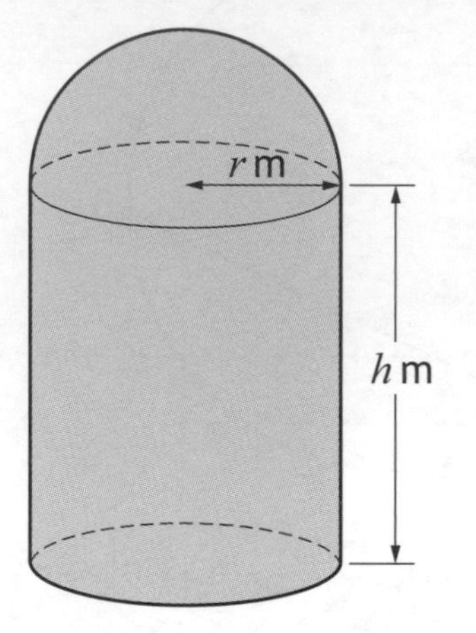

(a) Given that the curved surface area of a hemisphere of radius r is $2\pi r^2$ show that the surface area of metal needed to build the grain container is given by:

$$A = \frac{200}{r} + 3\pi r^2 \text{ square metres}$$ **3**

(b) Determine the value of r which minimises the amount of metal needed to build the container. **6**

MARKS

9. A sea-life visitor attraction has a new logo in the shape of a shark fin.

The outline of the logo can be represented by parts of

- the x axis
- the curve with equation $y = \cos(2x)$
- the curve with equation $y = \sin\left(\frac{3}{4}x - \frac{3}{2}\pi\right)$

as shown in the diagram.

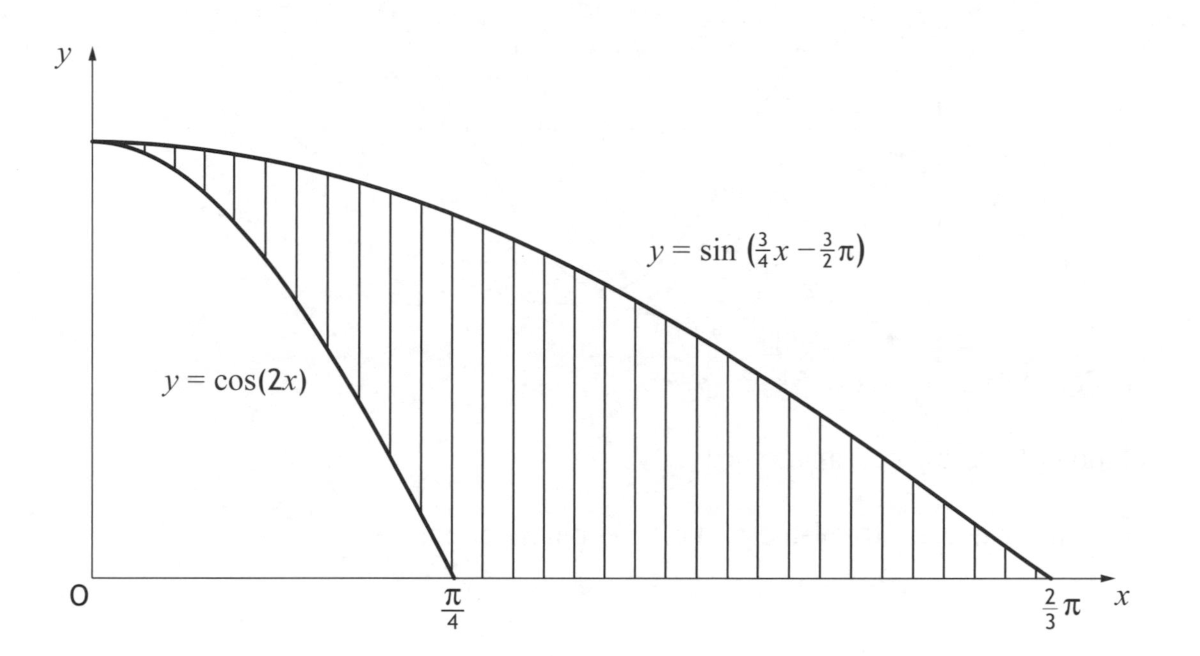

Calculate the shaded area. **6**

MARKS

10. Two sound sources produce the waves $y = \sin t$ and $y = \sqrt{3}\cos t$.

An investigation into the addition of these two waves produces the graph shown, with equation $y = k\cos(t - \text{a})$ for $0 \leq t \leq 2\pi$.

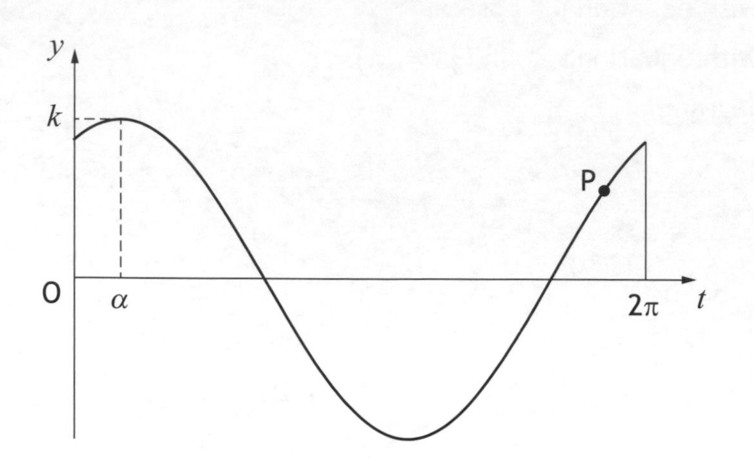

(a) Calculate the values of k and α. **4**

The point P has a y-coordinate of 1·2.

(b) Hence calculate the value of the t-coordinate of point P. **4**

[END OF SPECIMEN QUESTION PAPER]

HIGHER

Model Paper

Whilst this Model Paper has been specially commissioned by Hodder Gibson for use as practice for the Higher (for Curriculum for Excellence) exams, the key reference documents remain the SQA Specimen Paper 2014 and the SQA Past Papers 2015 and 2016.

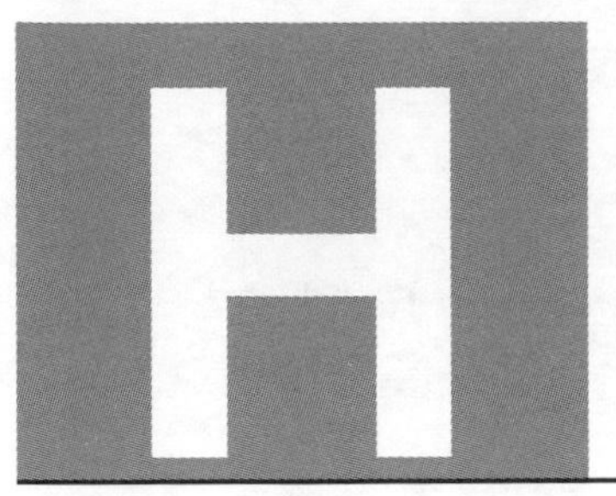

National
Qualifications
MODEL PAPER

Mathematics
Paper 1
(Non-Calculator)

Duration — 1 hour and 10 minutes

Total marks — 60

Attempt ALL questions.

You may NOT use a calculator.

Full credit will be given only to solutions which contain appropriate working.

State the units for your answer where appropriate.

Write your answers clearly in the answer booklet provided. In the answer booklet you must clearly identify the question number you are attempting.

Use **blue** or **black** ink.

Before leaving the examination room you must give your answer booklet to the Invigilator;
if you do not you may lose all the marks for this paper.

FORMULAE LIST

Circle:

The equation $x^2 + y^2 + 2gx + 2fy + c = 0$ represents a circle centre $(-g, -f)$ and radius $\sqrt{g^2 + f^2 - c}$.

The equation $(x - a)^2 + (y - b)^2 = r^2$ represents a circle centre (a, b) and radius r.

Scalar Product: $\mathbf{a.b} = |\mathbf{a}||\mathbf{b}| \cos \theta$, where θ is the angle between $\mathbf{a}$ and $\mathbf{b}$

or $\mathbf{a.b} = a_1b_1 + a_2b_2 + a_3b_3$ where $\mathbf{a} = \begin{pmatrix} a_1 \\ a_2 \\ a_3 \end{pmatrix}$ and $\mathbf{b} = \begin{pmatrix} b_1 \\ b_2 \\ b_3 \end{pmatrix}$

Trigonometric formulae:

$$\begin{aligned}
\sin(A \pm B) &= \sin A \cos B \pm \cos A \sin B \\
\cos(A \pm B) &= \cos A \cos B \mp \sin A \sin B \\
\sin 2A &= 2\sin A \cos A \\
\cos 2A &= \cos^2 A - \sin^2 A \\
&= 2\cos^2 A - 1 \\
&= 1 - 2\sin^2 A
\end{aligned}$$

Table of standard derivatives:

$f(x)$	$f'(x)$
$\sin ax$	$a \cos ax$
$\cos ax$	$-a \sin ax$

Table of standard integrals:

$f(x)$	$\int f(x)dx$
$\sin ax$	$-\frac{1}{a} \cos ax + C$
$\cos ax$	$\frac{1}{a} \sin ax + C$

Attempt ALL questions

Total marks — 60

MARKS

1. ABCD is a parallelogram. A, B and C have coordinates (2,3), (4,7) and (8, 11). Find the equation of DC. **3**

2. The point Q divides the line joining P(−1, −1, 0) to R(5, 2, −3) in the ratio 2:1. Find the coordinates of Q. **3**

3. Find the value of k such that the equation $kx^2 + kx + 6$, $k \neq 0$, has equal roots. **4**

4. Vectors **u** and **v** are defined by $\boldsymbol{u} = 3\mathbf{i} + 2\mathbf{j}$ and $\mathbf{v} = 2\mathbf{i} - 3\mathbf{j} + 4\mathbf{k}$. Determine whether or not **u** and **v** are perpendicular to each other. **2**

5. Find $\int \frac{4x^3 - 1}{x^2}\, dx$, $x \neq 0$. **4**

6. Functions $f(x) = \frac{1}{x-4}$ and $g(x) = 2x + 3$ are defined on suitable domains.

(a) Find an expression for $g^{-1}(x)$. **2**

(b) (i) Find an expression in simplest form for $h(x)$ where $h(x) = f(g)(x))$. **2**

(ii) Write down any restrictions on the domain of h. **1**

7. Explain why the equation $x^2 + y^2 + 2x + 3y + 5 = 0$ does **not** represent a circle. **2**

8. A ball is thrown vertically upwards. After t seconds its height is h metres, where $h = 1{\cdot}2 + 19{\cdot}6t - 4{\cdot}9t^2$.

(a) Find the speed of the ball after 1 second. **3**

(b) For how many seconds is the ball travelling upwards? **2**

9. (a) Express $7 - 2x - x^2$ in the form $a - (x + b)^2$. **2**

(b) State the maximum value of $7 - 2x - x^2$ and justify your answer. **2**

10. The diagram shows part of the graph of $y = \log_b(x + a)$. **MARKS**

Determine the values of a and b. **3**

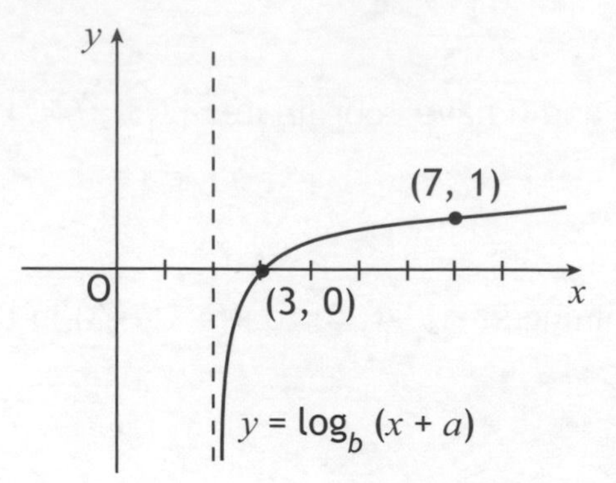

11. The graph of a function f intersects the x-axis at $(-a, 0)$ and $(e, 0)$ as shown.

There is a point of inflexion at $(0, b)$ and a maximum turning point at (c, d).

Sketch the graph of the derived function f' **3**

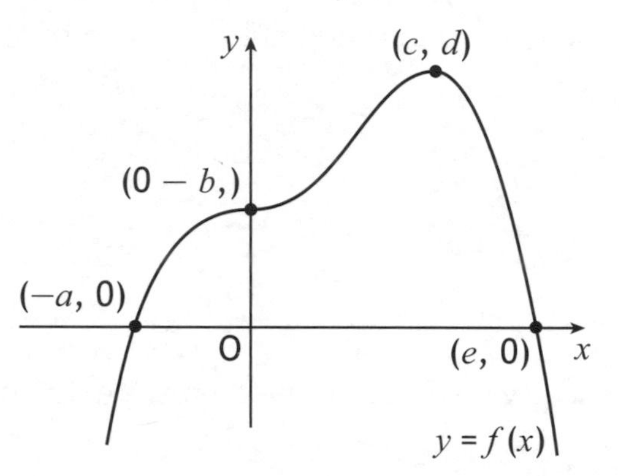

12. A function f is defined on the set of real numbers by $f(x) = x^3 - 3x + 2$.

(a) Find the coordinates of the stationary points on the curve $y = f(x)$ and determine their nature. **6**

(b) (i) Show that $(x - 1)$ is a factor of $x^3 - 3x + 2$. **3**

(ii) Hence or otherwise factorise $x^3 - 3x + 2$ fully. **2**

(c) State the coordinates of the points where the curve with equation $y = f(x)$ meets both the axes and hence sketch the curve. **4**

13. In the diagram **MARKS**

- angle DEC = angle CEB = $x°$
- angle CDE = angle BEA = $90°$
- CD = 1 unit
- DE = 3 units.

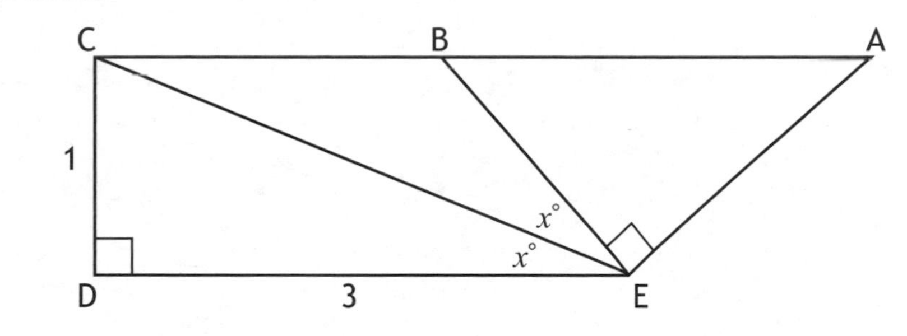

By writing angle DEA in terms of $x°$, find the exact value of cos(DEA). **7**

[END OF MODEL PAPER]

[BLANK PAGE]

DO NOT WRITE ON THIS PAGE

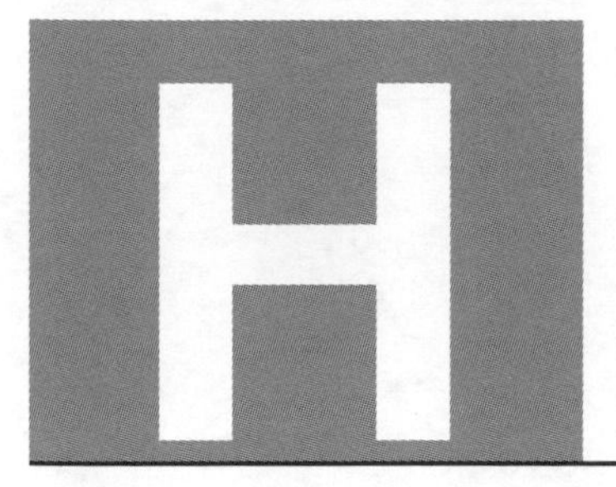

National
Qualifications
MODEL PAPER

Mathematics
Paper 2

Duration — 1 hour and 30 minutes

Total marks — 70

Attempt ALL questions.

You may use a calculator.

Full credit will be given only to solutions which contain appropriate working.

State the units for your answer where appropriate.

Write your answers clearly in the answer booklet provided. In the answer booklet you must clearly identify the question number you are attempting.

Use **blue** or **black** ink.

Before leaving the examination room you must give your answer booklet to the Invigilator; if you do not you may lose all the marks for this paper.

FORMULAE LIST

Circle:

The equation $x^2 + y^2 + 2gx + 2fy + c = 0$ represents a circle centre $(-g, -f)$ and radius $\sqrt{g^2 + f^2 - c}$.

The equation $(x - a)^2 + (y - b)^2 = r^2$ represents a circle centre (a, b) and radius r.

Scalar Product: $\mathbf{a.b} = |\mathbf{a}||\mathbf{b}| \cos \theta$, where θ is the angle between $\mathbf{a}$ and $\mathbf{b}$

or $\mathbf{a.b} = a_1b_1 + a_2b_2 + a_3b_3$ where $\mathbf{a} = \begin{pmatrix} a_1 \\ a_2 \\ a_3 \end{pmatrix}$ and $\mathbf{b} = \begin{pmatrix} b_1 \\ b_2 \\ b_3 \end{pmatrix}$

Trigonometric formulae:

$$\begin{aligned}
\sin(A \pm B) &= \sin A \cos B \pm \cos A \sin B \\
\cos(A \pm B) &= \cos A \cos B \mp \sin A \sin B \\
\sin 2A &= 2\sin A \cos A \\
\cos 2A &= \cos^2 A - \sin^2 A \\
&= 2\cos^2 A - 1 \\
&= 1 - 2\sin^2 A
\end{aligned}$$

Table of standard derivatives:

$f(x)$	$f'(x)$
$\sin ax$	$a \cos ax$
$\cos ax$	$-a \sin ax$

Table of standard integrals:

$f(x)$	$\int f(x)dx$
$\sin ax$	$-\frac{1}{a} \cos ax + C$
$\cos ax$	$\frac{1}{a} \sin ax + C$

Attempt ALL questions

Total marks — 70

MARKS

1. The vertices of triangle ABC are A(7, 9), B(−3, −1) and C(5, −5) as shown in the diagram.

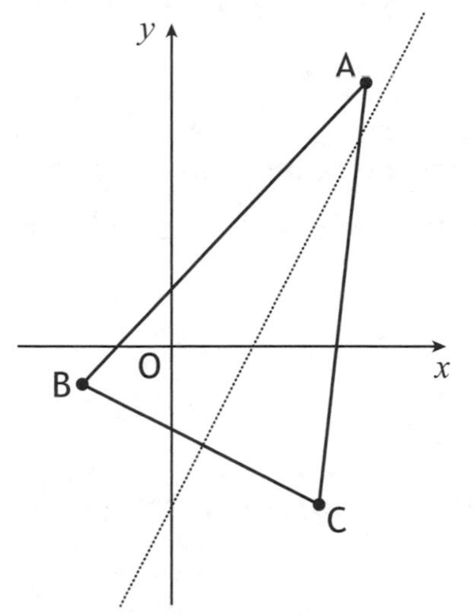

The broken line represents the perpendicular bisector of BC.

(a) Show that the equation of the perpendicular bisector of BC is $y = 2x - 5$. **4**

(b) Find the equation of the median from C. **3**

(c) Find the coordinates of the point of intersection of the perpendicular bisector of BC and the median from C. **3**

2. D,OABC is a square based pyramid as shown in the diagram.

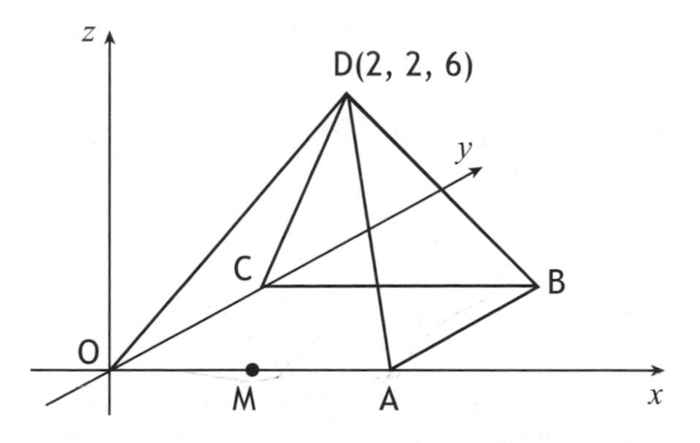

O is the origin, D is the point (2, 2, 6) and OA = 4 units.

M is the midpoint of OA.

(a) State the coordinates of B. **1**

(b) Express $\overrightarrow{DB}$ and $\overrightarrow{DM}$ in component form. **3**

(c) Find the size of angle BDM. **5**

MARKS

3. A man decides to plant a number of fast-growing trees as a boundary between his property and the property of his next door neighbour. He has been warned, however, by the local garden centre that, during any year, the trees are expected to increase in height by 0·5 metres. In response to this warning, he decides to trim 20% off the height of the trees at the start of any year.

(a) If he adopts the "20% pruning policy", to what height will he expect the trees to grow in the long run? **3**

(b) His neighbour is concerned that the trees are growing at an alarming rate and wants assurances that the trees will grow no taller than 2 metres.

What is the minimum percentage that the trees will need to be trimmed each year so as to meet this condition? **3**

4. The diagram shows the curve with equation $y = x^3 - x^2 - 4x + 4$ and the line with equation $y = 2x + 4$.

The curve and the line intersect at the points (–2, 0), (0, 4) and (3, 10).

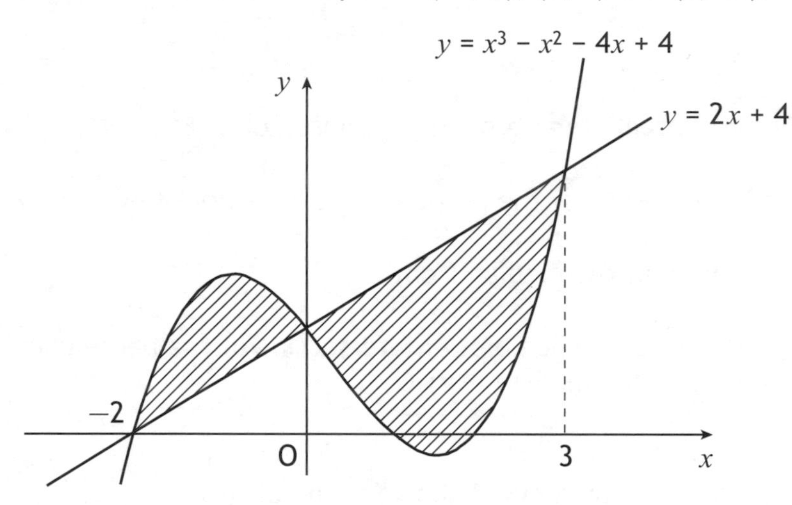

Calculate the total shaded area. **10**

5. (a) $12 \cos x° - 5 \sin x°$ can be expressed in the form $k \cos (x + a)°$, where $k > 0$ and $0 \le a < 360$.

Calculate the values of k and a. **4**

(b) (i) Hence state the maximum and minimum values of $12 \cos x° - 5 \sin x°$. **1**

(ii) Determine the values of x, in the interval $0 \le x < 360$ at which these maximum and minimum values occur. **2**

6. Find the value of $\int_0^2 \sin(4x + 1)dx$. **4**

MARKS

7. (a) (i) Show that the line with the equation $y = 3 - x$ is a tangent to the circle with the equation $x^2 + y^2 + 14x + 4y - 19 = 0$ **4**

(ii) Find the coordinates of the point of contact, P. **1**

(b) Relative to a suitable set of coordinate axes, the diagram below shows the circle from (a) and a second smaller circle with centre C.

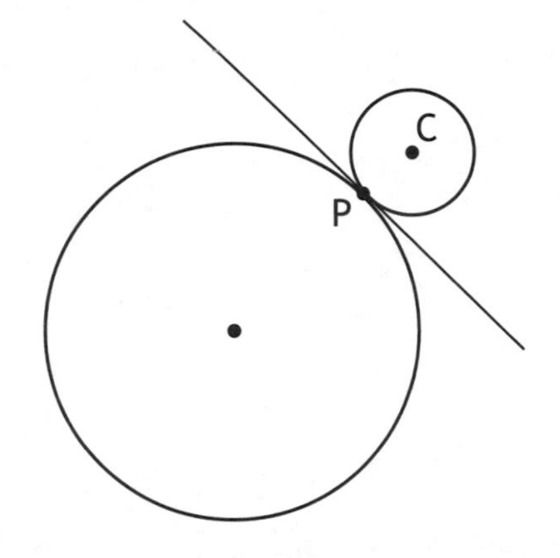

The line $y = 3 - x$ is a common tangent at the point P.

The radius of the larger circle is three times the radius of the smaller circle.

Find the equation of the smaller circle. **6**

8. The amount A_t micrograms of a certain radioactive substance remaining after t years decreases according to the formula $A_t = A_0 e^{-0 \cdot 002t}$, where A_0 is the amount present initially.

(a) If 600 micrograms are left after 1000 years, how many micrograms were present initially? **3**

(b) The half-life of a substance is the time taken for the amount to decrease to half of its initial amount. What is the half-life of this substance? **5**

9. Solve $2 \cos 2x - 5 \cos x - 4 = 0$ for $0 \leq x < 2\pi$ **5**

[END OF MODEL PAPER]

[BLANK PAGE]

DO NOT WRITE ON THIS PAGE

HIGHER

2015

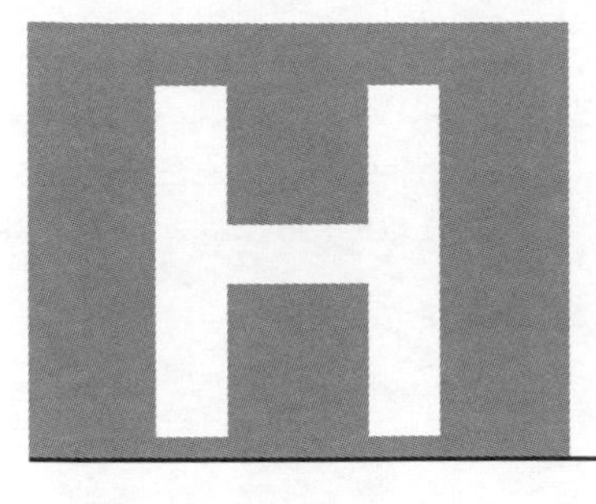

National
Qualifications
2015

X747/76/11

Mathematics Paper 1 (Non-Calculator)

WEDNESDAY, 20 MAY

9:00 AM — 10:10 AM

Total marks — 60

Attempt ALL questions.

You may NOT use a calculator.

Full credit will be given only to solutions which contain appropriate working.

State the units for your answer where appropriate.

Answers obtained by readings from scale drawings will not receive any credit.

Write your answers clearly in the spaces in the answer booklet provided. Additional space for answers is provided at the end of the answer booklet. If you use this space you must clearly identify the question number you are attempting.

Use **blue** or **black** ink.

Before leaving the examination room you must give your answer booklet to the Invigilator; if you do not, you may lose all the marks for this paper.

FORMULAE LIST

Circle:

The equation $x^2 + y^2 + 2gx + 2fy + c = 0$ represents a circle centre $(-g, -f)$ and radius $\sqrt{g^2 + f^2 - c}$.

The equation $(x - a)^2 + (y - b)^2 = r^2$ represents a circle centre (a, b) and radius r.

Scalar Product: $\mathbf{a}.\mathbf{b} = |\mathbf{a}||\mathbf{b}| \cos \theta$, where θ is the angle between $\mathbf{a}$ and $\mathbf{b}$

or $\mathbf{a}.\mathbf{b} = a_1b_1 + a_2b_2 + a_3b_3$ where $\mathbf{a} = \begin{pmatrix} a_1 \\ a_2 \\ a_3 \end{pmatrix}$ and $\mathbf{b} = \begin{pmatrix} b_1 \\ b_2 \\ b_3 \end{pmatrix}$.

Trigonometric formulae:

$$\begin{aligned}
\sin(A \pm B) &= \sin A \cos B \pm \cos A \sin B \\
\cos(A \pm B) &= \cos A \cos B \mp \sin A \sin B \\
\sin 2A &= 2 \sin A \cos A \\
\cos 2A &= \cos^2 A - \sin^2 A \\
&= 2\cos^2 A - 1 \\
&= 1 - 2\sin^2 A
\end{aligned}$$

Table of standard derivatives:

$f(x)$	$f'(x)$
$\sin ax$ $\cos ax$	$a \cos ax$ $-a \sin ax$

Table of standard integrals:

$f(x)$	$\int f(x)dx$
$\sin ax$ $\cos ax$	$-\frac{1}{a}\cos ax + c$ $\frac{1}{a}\sin ax + c$

MARKS

Attempt ALL questions

Total marks — 60

1. Vectors $\mathbf{u} = 8\mathbf{i} + 2\mathbf{j} - \mathbf{k}$ and $\mathbf{v} = -3\mathbf{i} + t\mathbf{j} - 6\mathbf{k}$ are perpendicular.
   Determine the value of t. **2**

2. Find the equation of the tangent to the curve $y = 2x^3 + 3$ at the point where $x = -2$. **4**

3. Show that $(x + 3)$ is a factor of $x^3 - 3x^2 - 10x + 24$ and hence factorise $x^3 - 3x^2 - 10x + 24$ fully. **4**

4. The diagram shows part of the graph of the function $y = p\cos qx + r$.

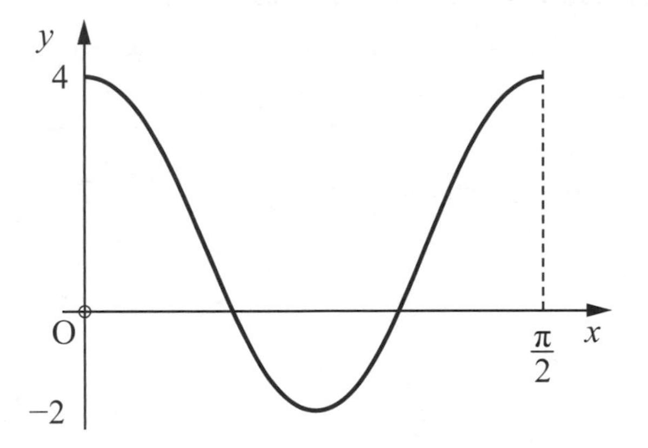

   Write down the values of p, q and r. **3**

5. A function g is defined on $\mathbb{R}$, the set of real numbers, by $g(x) = 6 - 2x$.

   (a) Determine an expression for $g^{-1}(x)$. **2**

   (b) Write down an expression for $g(g^{-1}(x))$. **1**

6. Evaluate $\log_6 12 + \frac{1}{3}\log_6 27$. **3**

7. A function f is defined on a suitable domain by $f(x) = \sqrt{x}\left(3x - \frac{2}{x\sqrt{x}}\right)$.

   Find $f'(4)$. **4**

[Turn over

MARKS

8. ABCD is a rectangle with sides of lengths x centimetres and $(x - 2)$ centimetres, as shown.

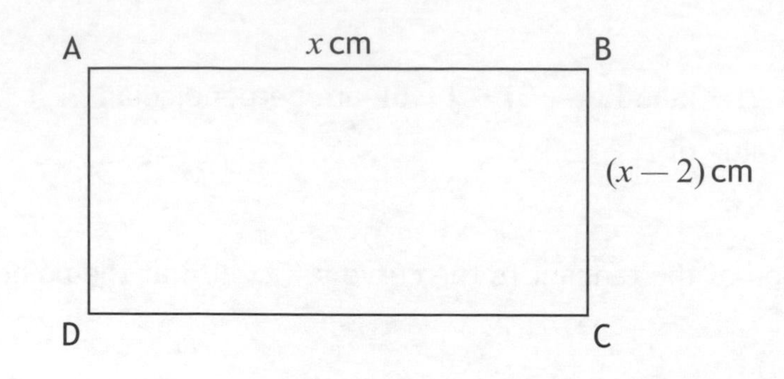

If the area of ABCD is less than $15\,\text{cm}^2$, determine the range of possible values of x. **4**

9. A, B and C are points such that AB is parallel to the line with equation $y + \sqrt{3}x = 0$ and BC makes an angle of $150°$ with the positive direction of the x-axis.

Are the points A, B and C collinear? **3**

10. Given that $\tan 2x = \frac{3}{4}$, $0 < x < \frac{\pi}{4}$, find the exact value of

(a) $\cos 2x$ **1**

(b) $\cos x$. **2**

11. T$(-2, -5)$ lies on the circumference of the circle with equation

$$(x + 8)^2 + (y + 2)^2 = 45.$$

(a) Find the equation of the tangent to the circle passing through T. **4**

(b) This tangent is also a tangent to a parabola with equation $y = -2x^2 + px + 1 - p$, where $p > 3$.

Determine the value of p. **6**

MARKS

12. The diagram shows part of the graph of $y = a\cos bx$.

The shaded area is $\frac{1}{2}$ unit^2.

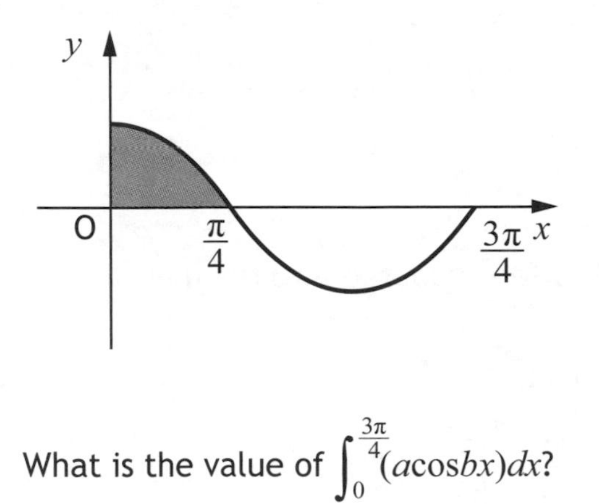

What is the value of $\int_0^{\frac{3\pi}{4}}(a\cos bx)dx$? **2**

13. The function $f(x) = 2^x + 3$ is defined on $\mathbb{R}$, the set of real numbers.

The graph with equation $y = f(x)$ passes through the point P$(1, b)$ and cuts the y-axis at Q as shown in the diagram.

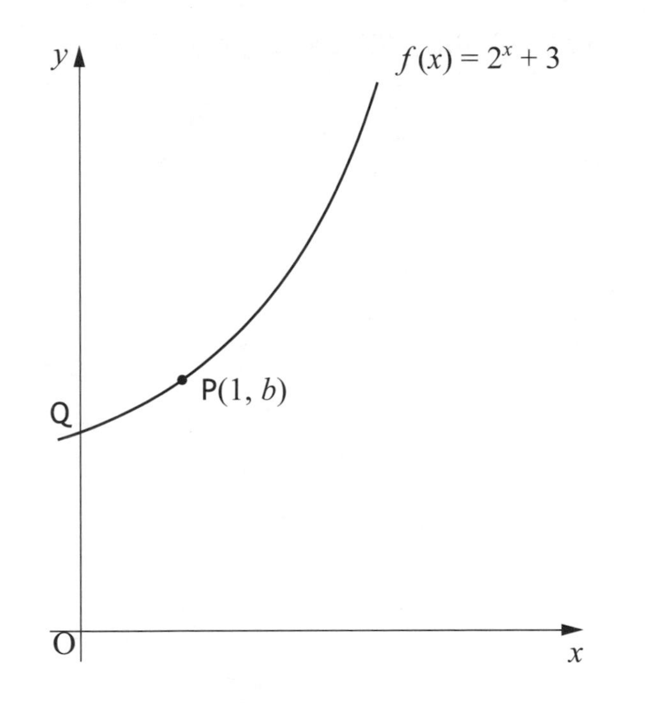

(a) What is the value of b? **1**

(b) (i) Copy the above diagram.

On the same diagram, sketch the graph with equation $y = f^{-1}(x)$. **1**

(ii) Write down the coordinates of the images of P and Q. **3**

(c) R $(3,11)$ also lies on the graph with equation $y = f(x)$.

Find the coordinates of the image of R on the graph with equation $y = 4 - f(x + 1)$. **2**

MARKS

14. The circle with equation $x^2 + y^2 - 12x - 10y + k = 0$ meets the coordinate axes at exactly three points.

What is the value of k? **2**

15. The rate of change of the temperature, T °C of a mug of coffee is given by

$$\frac{dT}{dt} = \frac{1}{25}t - k \text{ , } 0 \leq t \leq 50$$

- t is the elapsed time, in minutes, after the coffee is poured into the mug
- k is a constant
- initially, the temperature of the coffee is 100 °C
- 10 minutes later the temperature has fallen to 82 °C.

Express T in terms of t. **6**

[END OF QUESTION PAPER]

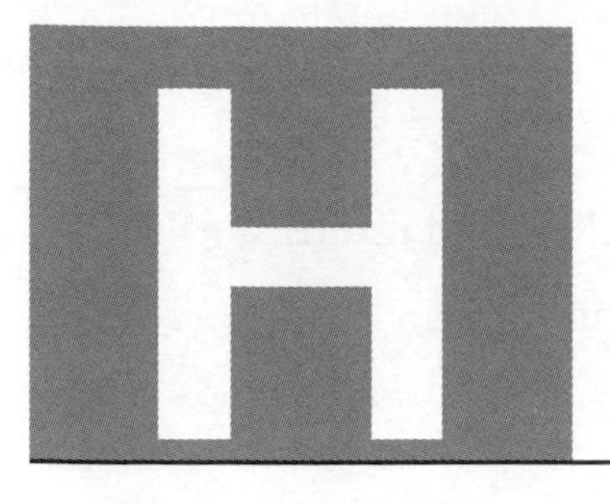

National
Qualifications
2015

X747/76/12

Mathematics
Paper 2

WEDNESDAY, 20 MAY
10:30 AM — 12:00 NOON

Total marks — 70

Attempt ALL questions.

You may use a calculator

Full credit will be given only to solutions which contain appropriate working.

State the units for your answer where appropriate.

Answers obtained by readings from scale drawings will not receive any credit.

Write your answers clearly in the spaces in the answer booklet provided. Additional space for answers is provided at the end of the answer booklet. If you use this space you must clearly identify the question number you are attempting.

Use **blue** or **black** ink.

Before leaving the examination room you must give your answer booklet to the Invigilator; if you do not, you may lose all the marks for this paper.

FORMULAE LIST

Circle:

The equation $x^2 + y^2 + 2gx + 2fy + c = 0$ represents a circle centre $(-g, -f)$ and radius $\sqrt{g^2 + f^2 - c}$.

The equation $(x - a)^2 + (y - b)^2 = r^2$ represents a circle centre (a, b) and radius r.

Scalar Product: $\mathbf{a}.\mathbf{b} = |\mathbf{a}||\mathbf{b}| \cos \theta$, where θ is the angle between $\mathbf{a}$ and $\mathbf{b}$

or $\mathbf{a}.\mathbf{b} = a_1b_1 + a_2b_2 + a_3b_3$ where $\mathbf{a} = \begin{pmatrix} a_1 \\ a_2 \\ a_3 \end{pmatrix}$ and $\mathbf{b} = \begin{pmatrix} b_1 \\ b_2 \\ b_3 \end{pmatrix}$.

Trigonometric formulae:

$$\sin(A \pm B) = \sin A \cos B \pm \cos A \sin B$$
$$\cos(A \pm B) = \cos A \cos B \mp \sin A \sin B$$
$$\sin 2A = 2 \sin A \cos A$$
$$\cos 2A = \cos^2 A - \sin^2 A$$
$$= 2\cos^2 A - 1$$
$$= 1 - 2\sin^2 A$$

Table of standard derivatives:

$f(x)$	$f'(x)$
$\sin ax$	$a \cos ax$
$\cos ax$	$-a \sin ax$

Table of standard integrals:

$f(x)$	$\int f(x)dx$
$\sin ax$	$-\frac{1}{a} \cos ax + c$
$\cos ax$	$\frac{1}{a} \sin ax + c$

MARKS

Attempt ALL questions

Total marks — 70

1. The vertices of triangle ABC are A$(-5, 7)$, B$(-1, -5)$ and C$(13, 3)$ as shown in the diagram.

   The broken line represents the altitude from C.

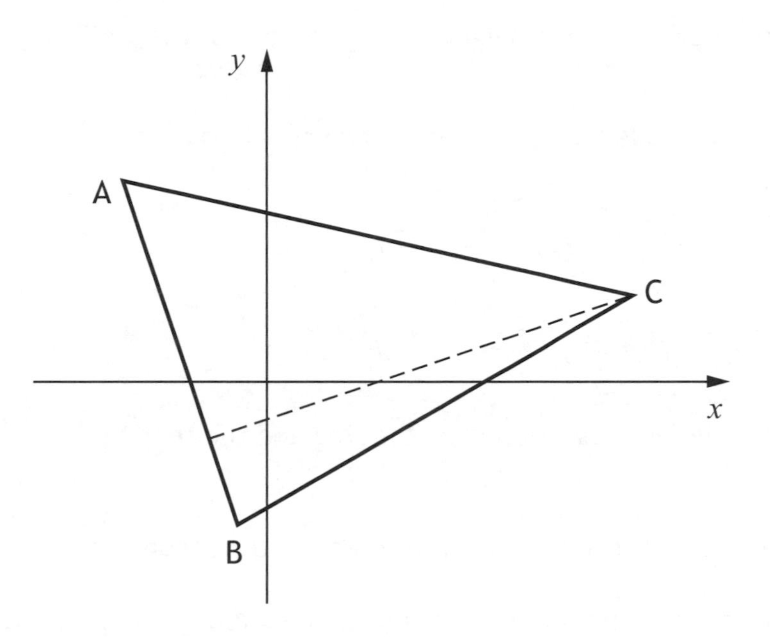

   (a) Show that the equation of the altitude from C is $x - 3y = 4$. **4**

   (b) Find the equation of the median from B. **3**

   (c) Find the coordinates of the point of intersection of the altitude from C and the median from B. **2**

2. Functions f and g are defined on suitable domains by

   $$f(x) = 10 + x \text{ and } g(x) = (1 + x)(3 - x) + 2.$$

   (a) Find an expression for $f(g(x))$. **2**

   (b) Express $f(g(x))$ in the form $p(x + q)^2 + r$. **3**

   (c) Another function h is given by $h(x) = \dfrac{1}{f(g(x))}$.

   What values of x cannot be in the domain of h? **2**

[Turn over

MARKS

3. A version of the following problem first appeared in print in the 16th Century.

A frog and a toad fall to the bottom of a well that is 50 feet deep.

Each day, the frog climbs 32 feet and then rests overnight. During the night, it slides down $\frac{2}{3}$ of its height above the floor of the well.

The toad climbs 13 feet each day before resting.

Overnight, it slides down $\frac{1}{4}$ of its height above the floor of the well.

Their progress can be modelled by the recurrence relations:

- $f_{n+1} = \frac{1}{3}f_n + 32, \qquad f_1 = 32$
- $t_{n+1} = \frac{3}{4}t_n + 13, \qquad t_1 = 13$

where f_n and t_n are the heights reached by the frog and the toad at the end of the nth day after falling in.

(a) Calculate t_2, the height of the toad at the end of the second day. **1**

(b) Determine whether or not either of them will eventually escape from the well. **5**

MARKS

4. A wall plaque is to be made to commemorate the 150th anniversary of the publication of "*Alice's Adventures in Wonderland*".

The edges of the wall plaque can be modelled by parts of the graphs of four quadratic functions as shown in the sketch.

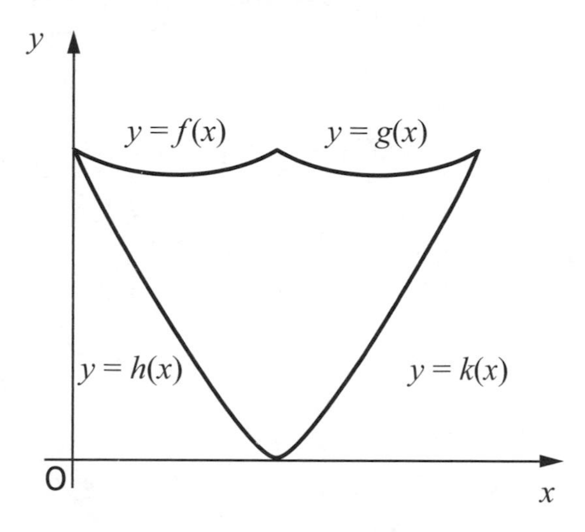

- $f(x)=\frac{1}{4}x^2-\frac{1}{2}x+3$
- $g(x)=\frac{1}{4}x^2-\frac{3}{2}x+5$
- $h(x)=\frac{3}{8}x^2-\frac{9}{4}x+3$
- $k(x)=\frac{3}{8}x^2-\frac{3}{4}x$

(a) Find the x-coordinate of the point of intersection of the graphs with equations $y=f(x)$ and $y=g(x)$. **2**

The graphs of the functions $f(x)$ and $h(x)$ intersect on the y-axis.

The plaque has a vertical line of symmetry.

(b) Calculate the area of the wall plaque. **7**

[Turn over

MARKS

5. Circle C_1 has equation $x^2 + y^2 + 6x + 10y + 9 = 0$.

The centre of circle C_2 is $(9, 11)$.

Circles C_1 and C_2 touch externally.

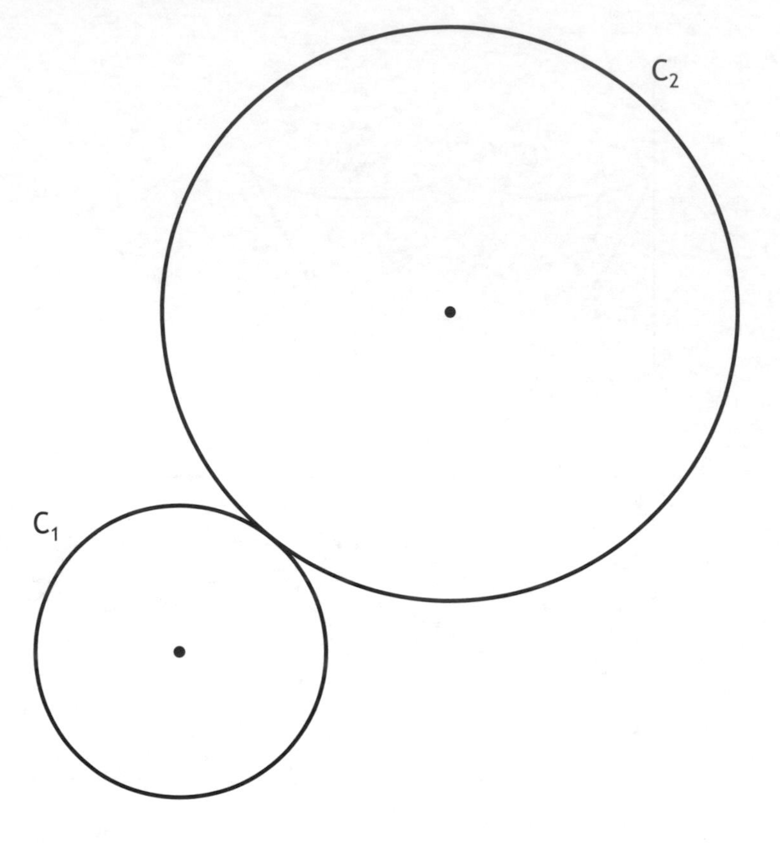

(a) Determine the radius of C_2. **4**

A third circle, C_3, is drawn such that:

- both C_1 and C_2 touch C_3 internally
- the centres of C_1, C_2 and C_3 are collinear.

(b) Determine the equation of C_3. **4**

MARKS

6. Vectors $\mathbf{p}$, $\mathbf{q}$ and $\mathbf{r}$ are represented on the diagram as shown.

- BCDE is a parallelogram
- ABE is an equilateral triangle
- $|\mathbf{p}| = 3$
- Angle ABC $= 90°$

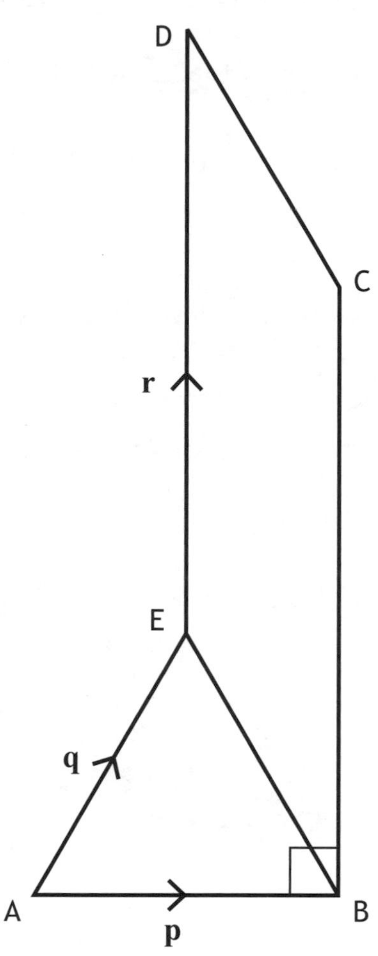

(a) Evaluate $\mathbf{p}.(\mathbf{q}+\mathbf{r})$. 3

(b) Express $\overrightarrow{EC}$ in terms of $\mathbf{p}$, $\mathbf{q}$ and $\mathbf{r}$. 1

(c) Given that $\overrightarrow{AE}.\overrightarrow{EC} = 9\sqrt{3} - \frac{9}{2}$, find $|\mathbf{r}|$. 3

[Turn over

MARKS

7. (a) Find $\int (3\cos 2x + 1)\,dx$. **2**

(b) Show that $3\cos 2x + 1 = 4\cos^2 x - 2\sin^2 x$. **2**

(c) Hence, or otherwise, find $\int \left(\sin^2 x - 2\cos^2 x\right) dx$. **2**

8. A crocodile is stalking prey located 20 metres further upstream on the opposite bank of a river.

Crocodiles travel at different speeds on land and in water.

The time taken for the crocodile to reach its prey can be minimised if it swims to a particular point, P, x metres upstream on the other side of the river as shown in the diagram.

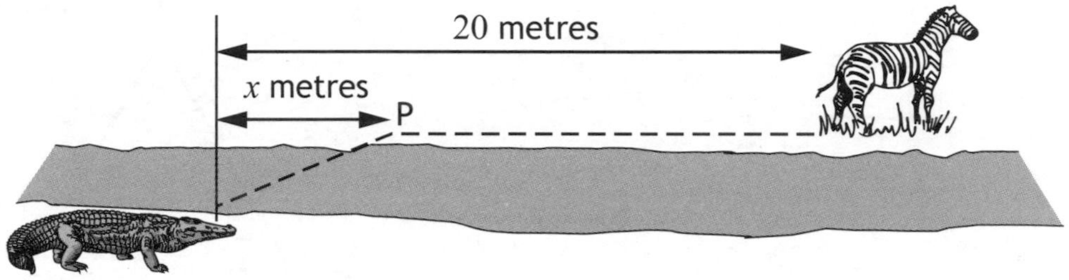

20 metres

x metres

P

The time taken, T, measured in tenths of a second, is given by

$$T(x) = 5\sqrt{36 + x^2} + 4(20 - x)$$

(a) (i) Calculate the time taken if the crocodile does not travel on land. **1**

(ii) Calculate the time taken if the crocodile swims the shortest distance possible. **1**

(b) Between these two extremes there is one value of x which minimises the time taken. Find this value of x and hence calculate the minimum possible time. **8**

MARKS

9. The blades of a wind turbine are turning at a steady rate.

The height, h metres, of the tip of one of the blades above the ground at time, t seconds, is given by the formula

$$h = 36\sin(1\cdot5t) - 15\cos(1\cdot5t) + 65.$$

Express $36\sin(1\cdot5t) - 15\cos(1\cdot5t)$ in the form

$k\sin(1\cdot5t - a)$, where $k > 0$ and $0 < a < \frac{\pi}{2}$,

and hence find the **two** values of t for which the tip of this blade is at a height of 100 metres above the ground during the first turn. **8**

[END OF QUESTION PAPER]

[BLANK PAGE]

DO NOT WRITE ON THIS PAGE

HIGHER

2016

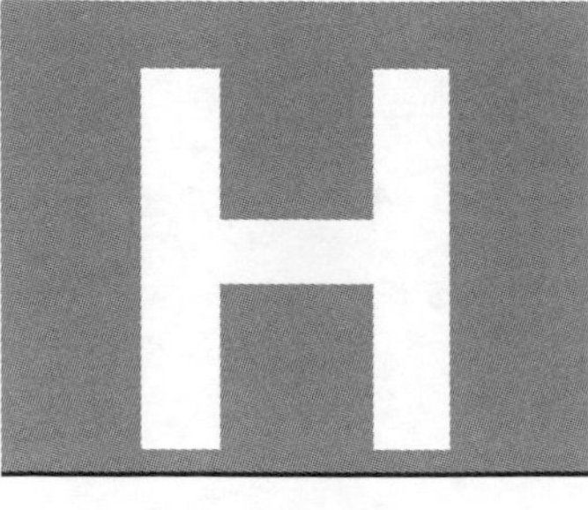

National
Qualifications
2016

X747/76/11

Mathematics Paper 1 (Non-Calculator)

THURSDAY, 12 MAY

9:00 AM — 10:10 AM

Total marks — 60

Attempt ALL questions.

You may NOT use a calculator.

Full credit will be given only to solutions which contain appropriate working.

State the units for your answer where appropriate.

Answers obtained by readings from scale drawings will not receive any credit.

Write your answers clearly in the spaces provided in the answer booklet. The size of the space provided for an answer should not be taken as an indication of how much to write. It is not necessary to use all the space.

Additional space for answers is provided at the end of the answer booklet. If you use this space **you must clearly identify the question number** you are attempting.

Use **blue** or **black** ink.

Before leaving the examination room you must give your answer booklet to the Invigilator; if you do not, you may lose all the marks for this paper.

FORMULAE LIST

Circle:

The equation $x^2 + y^2 + 2gx + 2fy + c = 0$ represents a circle centre $(-g, -f)$ and radius $\sqrt{g^2 + f^2 - c}$.

The equation $(x - a)^2 + (y - b)^2 = r^2$ represents a circle centre (a, b) and radius r.

Scalar Product: $\mathbf{a.b} = |\mathbf{a}||\mathbf{b}| \cos \theta$, where θ is the angle between $\mathbf{a}$ and $\mathbf{b}$

or $\mathbf{a.b} = a_1b_1 + a_2b_2 + a_3b_3$ where $\mathbf{a} = \begin{pmatrix} a_1 \\ a_2 \\ a_3 \end{pmatrix}$ and $\mathbf{b} = \begin{pmatrix} b_1 \\ b_2 \\ b_3 \end{pmatrix}$.

Trigonometric formulae:

$$\sin(A \pm B) = \sin A \cos B \pm \cos A \sin B$$
$$\cos(A \pm B) = \cos A \cos B \mp \sin A \sin B$$
$$\sin 2A = 2 \sin A \cos A$$
$$\cos 2A = \cos^2 A - \sin^2 A$$
$$= 2\cos^2 A - 1$$
$$= 1 - 2\sin^2 A$$

Table of standard derivatives:

$f(x)$	$f'(x)$
$\sin ax$	$a \cos ax$
$\cos ax$	$-a \sin ax$

Table of standard integrals:

$f(x)$	$\int f(x)dx$
$\sin ax$	$-\frac{1}{a}\cos ax + c$
$\cos ax$	$\frac{1}{a}\sin ax + c$

MARKS

Attempt ALL questions

Total marks — 60

1. Find the equation of the line passing through the point $(-2, 3)$ which is parallel to the line with equation $y + 4x = 7$. **2**

2. Given that $y = 12x^3 + 8\sqrt{x}$, where $x > 0$, find $\frac{dy}{dx}$. **3**

3. A sequence is defined by the recurrence relation $u_{n+1} = \frac{1}{3}u_n + 10$ with $u_3 = 6$.

   (a) Find the value of u_4. **1**

   (b) Explain why this sequence approaches a limit as $n \to \infty$. **1**

   (c) Calculate this limit. **2**

4. A and B are the points $(-7, 3)$ and $(1, 5)$.

   AB is a diameter of a circle.

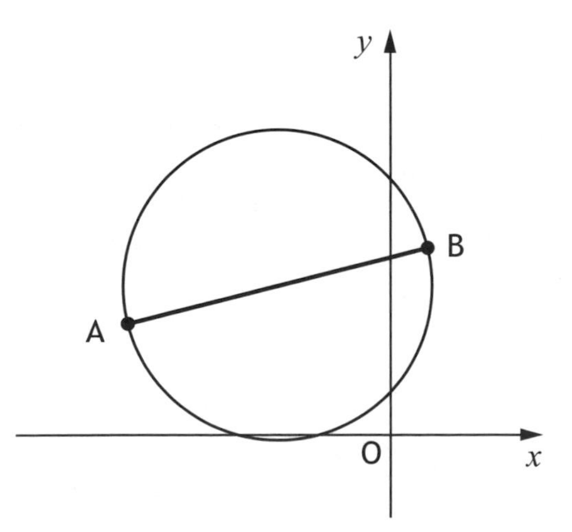

Find the equation of this circle. **3**

MARKS

5. Find $\int 8\cos(4x+1)\,dx$. **2**

6. Functions f and g are defined on $\mathbb{R}$, the set of real numbers.
The inverse functions f^{-1} and g^{-1} both exist.

(a) Given $f(x)=3x+5$, find $f^{-1}(x)$. **3**

(b) If $g(2)=7$, write down the value of $g^{-1}(7)$. **1**

7. Three vectors can be expressed as follows:

$\overrightarrow{FG} = -2\mathbf{i} - 6\mathbf{j} + 3\mathbf{k}$

$\overrightarrow{GH} = 3\mathbf{i} + 9\mathbf{j} - 7\mathbf{k}$

$\overrightarrow{EH} = 2\mathbf{i} + 3\mathbf{j} + \mathbf{k}$

(a) Find $\overrightarrow{FH}$. **2**

(b) Hence, or otherwise, find $\overrightarrow{FE}$. **2**

8. Show that the line with equation $y=3x-5$ is a tangent to the circle with equation $x^2+y^2+2x-4y-5=0$ and find the coordinates of the point of contact. **5**

MARKS

9. (a) Find the x-coordinates of the stationary points on the graph with equation $y = f(x)$, where $f(x) = x^3 + 3x^2 - 24x$. **4**

(b) Hence determine the range of values of x for which the function f is strictly increasing. **2**

10. The diagram below shows the graph of the function $f(x) = \log_4 x$, where $x > 0$.

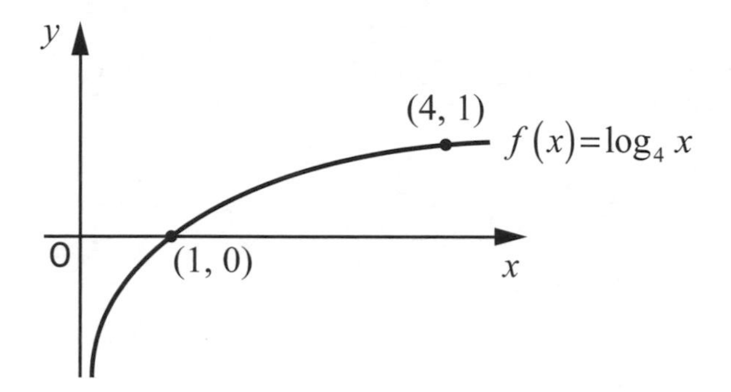

The inverse function, f^{-1}, exists.

On the diagram in your answer booklet, sketch the graph of the inverse function. **2**

11. (a) A and C are the points $(1, 3, -2)$ and $(4, -3, 4)$ respectively.

Point B divides AC in the ratio $1 : 2$.

Find the coordinates of B. **2**

(b) $k\overrightarrow{AC}$ is a vector of magnitude 1, where $k > 0$.

Determine the value of k. **3**

MARKS

12. The functions f and g are defined on $\mathbb{R}$, the set of real numbers by $f(x) = 2x^2 - 4x + 5$ and $g(x) = 3 - x$.

(a) Given $h(x) = f(g(x))$, show that $h(x) = 2x^2 - 8x + 11$. **2**

(b) Express $h(x)$ in the form $p(x+q)^2 + r$. **3**

13. Triangle ABD is right-angled at B with angles BAC $= p$ and BAD $= q$ and lengths as shown in the diagram below.

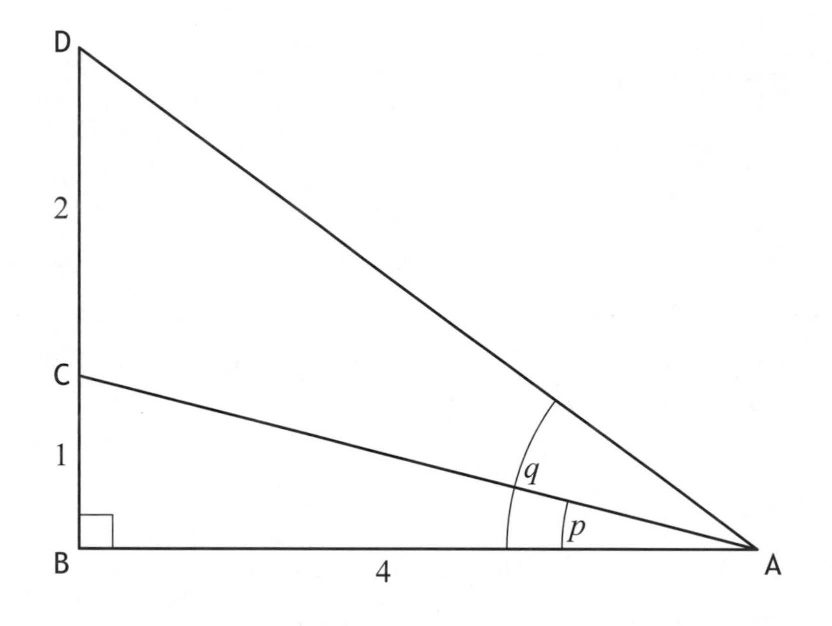

Show that the exact value of $\cos(q - p)$ is $\dfrac{19\sqrt{17}}{85}$. **5**

MARKS

14. (a) Evaluate $\log_5 25$. **1**

(b) Hence solve $\log_4 x + \log_4 (x-6) = \log_5 25$, where $x > 6$. **5**

15. The diagram below shows the graph with equation $y = f(x)$, where $f(x) = k(x-a)(x-b)^2$.

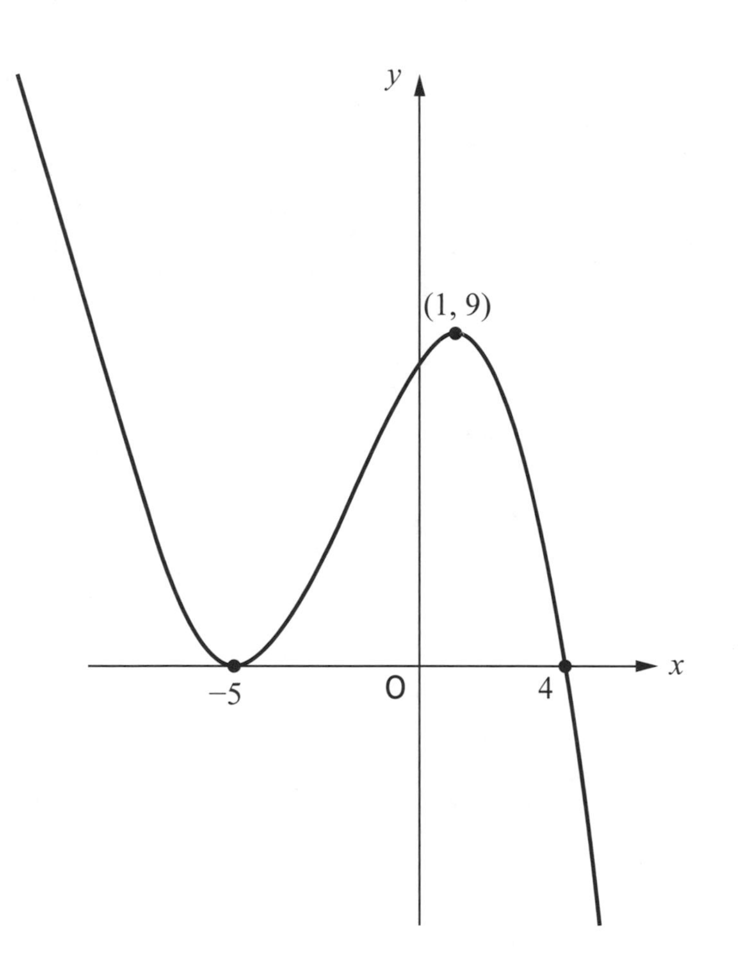

(a) Find the values of a, b and k. **3**

(b) For the function $g(x) = f(x) - d$, where d is positive, determine the range of values of d for which $g(x)$ has exactly one real root. **1**

[END OF QUESTION PAPER]

[BLANK PAGE]

DO NOT WRITE ON THIS PAGE

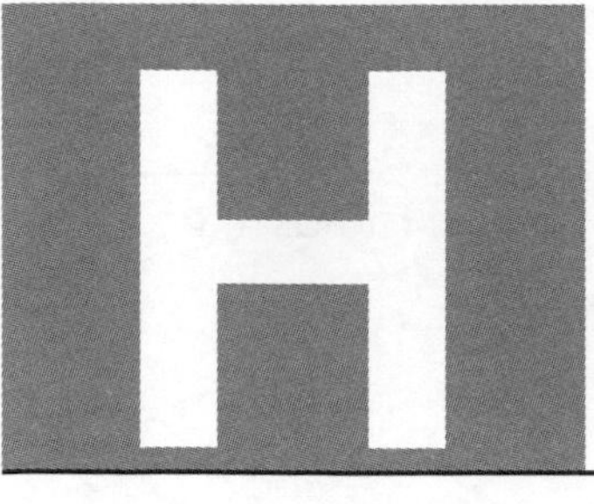

National
Qualifications
2016

X747/76/12

Mathematics Paper 2

THURSDAY, 12 MAY

10:30 AM — 12:00 NOON

Total marks — 70

Attempt ALL questions.

You may use a calculator.

Full credit will be given only to solutions which contain appropriate working.

State the units for your answer where appropriate.

Answers obtained by readings from scale drawings will not receive any credit.

Write your answers clearly in the spaces provided in the answer booklet. The size of the space provided for an answer should not be taken as an indication of how much to write. It is not necessary to use all the space.

Additional space for answers is provided at the end of the answer booklet. If you use this space **you must clearly identify the question number** you are attempting.

Use **blue** or **black** ink.

Before leaving the examination room you must give your answer booklet to the Invigilator; if you do not, you may lose all the marks for this paper.

FORMULAE LIST

Circle:

The equation $x^2 + y^2 + 2gx + 2fy + c = 0$ represents a circle centre $(-g, -f)$ and radius $\sqrt{g^2 + f^2 - c}$.

The equation $(x - a)^2 + (y - b)^2 = r^2$ represents a circle centre (a, b) and radius r.

Scalar Product: $\mathbf{a.b} = |\mathbf{a}||\mathbf{b}| \cos \theta$, where θ is the angle between $\mathbf{a}$ and $\mathbf{b}$

or $\mathbf{a.b} = a_1b_1 + a_2b_2 + a_3b_3$ where $\mathbf{a} = \begin{pmatrix} a_1 \\ a_2 \\ a_3 \end{pmatrix}$ and $\mathbf{b} = \begin{pmatrix} b_1 \\ b_2 \\ b_3 \end{pmatrix}$.

Trigonometric formulae:

$$\begin{aligned}
\sin(\mathrm{A} \pm \mathrm{B}) &= \sin \mathrm{A} \cos \mathrm{B} \pm \cos \mathrm{A} \sin \mathrm{B} \\
\cos(\mathrm{A} \pm \mathrm{B}) &= \cos \mathrm{A} \cos \mathrm{B} \mp \sin \mathrm{A} \sin \mathrm{B} \\
\sin 2\mathrm{A} &= 2 \sin \mathrm{A} \cos \mathrm{A} \\
\cos 2\mathrm{A} &= \cos^2 \mathrm{A} - \sin^2 \mathrm{A} \\
&= 2\cos^2 \mathrm{A} - 1 \\
&= 1 - 2\sin^2 \mathrm{A}
\end{aligned}$$

Table of standard derivatives:

$f(x)$	$f'(x)$
$\sin ax$	$a \cos ax$
$\cos ax$	$-a \sin ax$

Table of standard integrals:

$f(x)$	$\int f(x)dx$
$\sin ax$	$-\frac{1}{a}\cos ax + c$
$\cos ax$	$\frac{1}{a}\sin ax + c$

MARKS

Attempt ALL questions

Total marks — 70

1. PQR is a triangle with vertices P$(0,-4)$, Q$(-6,2)$ and R$(10,6)$.

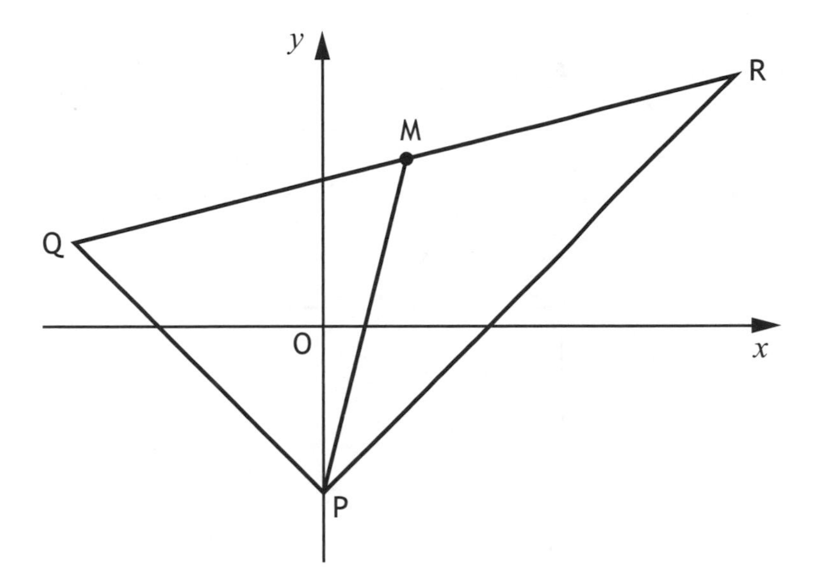

(a) (i) State the coordinates of M, the midpoint of QR. **1**

(ii) Hence find the equation of PM, the median through P. **2**

(b) Find the equation of the line, L, passing through M and perpendicular to PR. **3**

(c) Show that line L passes through the midpoint of PR. **3**

2. Find the range of values for p such that $x^2 - 2x + 3 - p = 0$ has no real roots. **3**

[Turn over

MARKS

3. (a) (i) Show that $(x+1)$ is a factor of $2x^3 - 9x^2 + 3x + 14$. **2**

(ii) Hence solve the equation $2x^3 - 9x^2 + 3x + 14 = 0$. **3**

(b) The diagram below shows the graph with equation $y = 2x^3 - 9x^2 + 3x + 14$.

The curve cuts the x-axis at A, B and C.

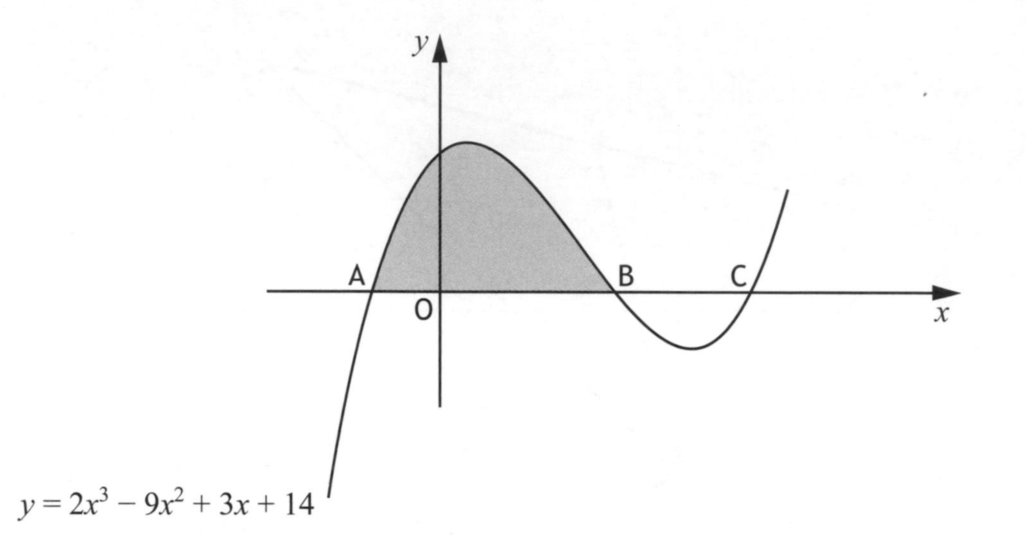

(i) Write down the coordinates of the points A and B. **1**

(ii) Hence calculate the shaded area in the diagram. **4**

4. Circles C_1 and C_2 have equations $(x+5)^2 + (y-6)^2 = 9$ and $x^2 + y^2 - 6x - 16 = 0$ respectively.

(a) Write down the centres and radii of C_1 and C_2. **4**

(b) Show that C_1 and C_2 do not intersect. **3**

MARKS

5. The picture shows a model of a water molecule.

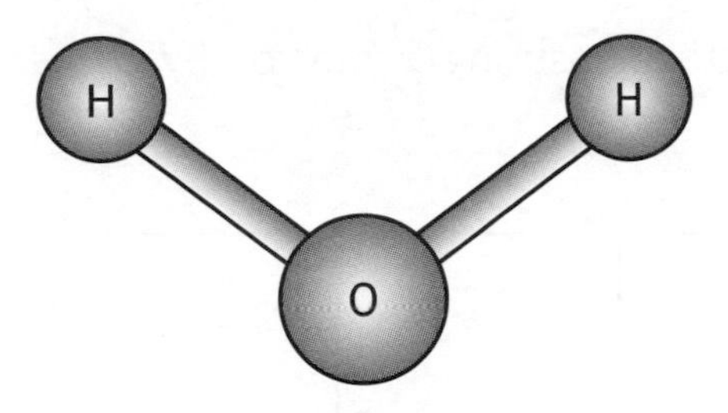

Relative to suitable coordinate axes, the oxygen atom is positioned at point A$(-2, 2, 5)$.

The two hydrogen atoms are positioned at points B$(-10, 18, 7)$ and C$(-4, -6, 21)$ as shown in the diagram below.

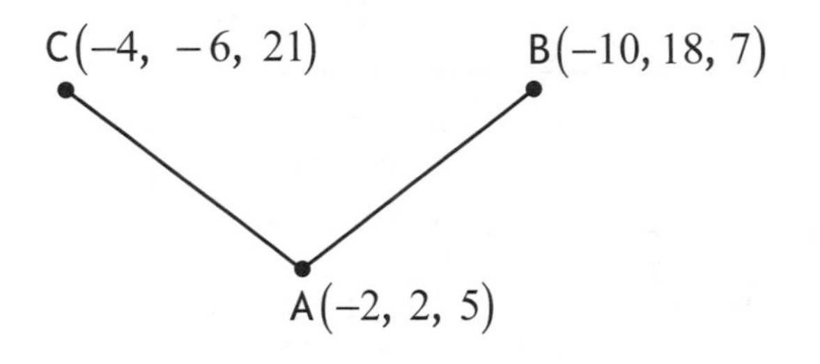

(a) Express $\overrightarrow{AB}$ and $\overrightarrow{AC}$ in component form. **2**

(b) Hence, or otherwise, find the size of angle BAC. **4**

6. Scientists are studying the growth of a strain of bacteria. The number of bacteria present is given by the formula

$$B(t) = 200e^{0 \cdot 107t},$$

where t represents the number of hours since the study began.

(a) State the number of bacteria present at the start of the study. **1**

(b) Calculate the time taken for the number of bacteria to double. **4**

[Turn over

MARKS

7. A council is setting aside an area of land to create six fenced plots where local residents can grow their own food.

Each plot will be a rectangle measuring x metres by y metres as shown in the diagram.

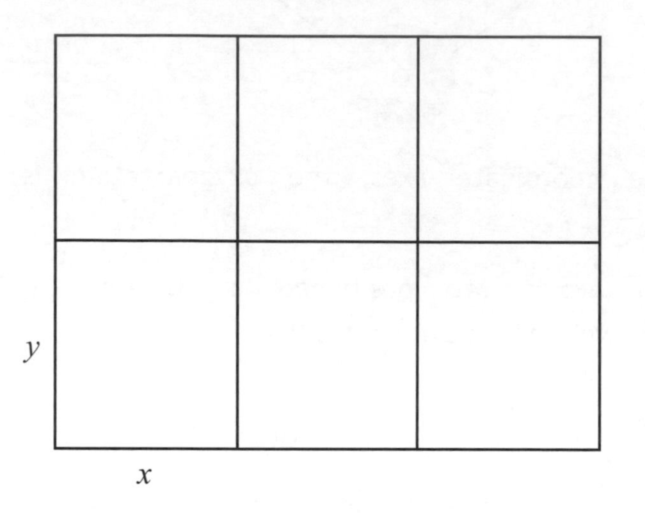

(a) The area of land being set aside is $108\,\text{m}^2$.

Show that the total length of fencing, L metres, is given by

$$L(x) = 9x + \frac{144}{x}.$$ **3**

(b) Find the value of x that minimises the length of fencing required. **6**

MARKS

8. (a) Express $5\cos x - 2\sin x$ in the form $k\cos(x + a)$, where $k > 0$ and $0 < a < 2\pi$. **4**

(b) The diagram shows a sketch of part of the graph of $y = 10 + 5\cos x - 2\sin x$ and the line with equation $y = 12$.

The line cuts the curve at the points P and Q.

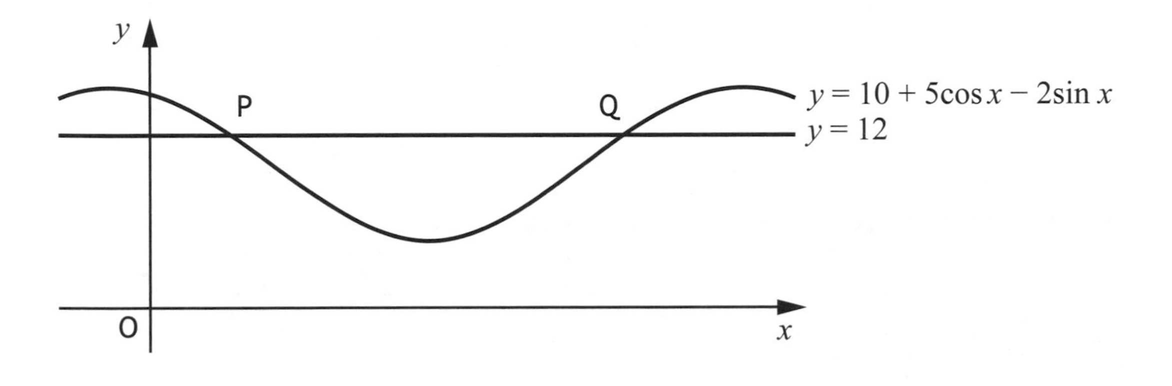

Find the x-coordinates of P and Q. **4**

9. For a function f, defined on a suitable domain, it is known that:

- $f'(x) = \dfrac{2x+1}{\sqrt{x}}$
- $f(9) = 40$

Express $f(x)$ in terms of x. **4**

[Turn over for next question

MARKS

10. (a) Given that $y=\left(x^2+7\right)^{\frac{1}{2}}$, find $\frac{dy}{dx}$. **2**

(b) Hence find $\int \frac{4x}{\sqrt{x^2+7}}\,dx$. **1**

11. (a) Show that $\sin 2x \tan x = 1 - \cos 2x$, where $\frac{\pi}{2} < x < \frac{3\pi}{2}$. **4**

(b) Given that $f(x)=\sin 2x \tan x$, find $f'(x)$. **2**

[END OF QUESTION PAPER]

HIGHER

Answers

ANSWERS FOR

SQA & HODDER GIBSON HIGHER MATHEMATICS 2016

HIGHER MATHEMATICS 2014 SPECIMEN QUESTION PAPER

Paper 1 (Non-Calculator)

Question			Generic Scheme. Give one mark for each •	Illustrative Scheme	Max mark
1.			Ans: $\frac{3}{4}x^2 - \frac{1}{2}x^{-1} + C$		4
			•1 preparation for integration	•1 $\frac{3}{2}x + \frac{1}{2}x^{-2}$	
			•2 correct integration of first term	•2 $\frac{3}{2} \times \frac{x^2}{2} + \ldots$	
			•3 correct integration of second term	•3 $\ldots + \frac{1}{2} \times \frac{x^{-1}}{-1}$	
			•4 includes constant of integration	•4 $\frac{3}{4}x^2 - \frac{1}{2}x^{-1} + C$	
2.			Ans: $(-1,0),(0,4),(3,16)$		5
			•1 sets equation of curve equal to equation of line	•1 $x^3 - 2x^2 + x + 4 = 4x + 4$	
			•2 equates to zero	•2 $x^3 - 2x^2 - 3x = 0$	
			•3 factorises fully	•3 $x(x+1)(x-3) = 0$	
			•4 calculates x-coordinates	•4 $x = 0, x = -1, x = 3$	
			•5 calculates y-coordinates	•5 $(0,4),(-1,0),(3,16)$	
3.			Ans: S(5, 25, −2)		5
			•1 find coordinate of Q or component vector $\mathbf{q}$	•1 $\mathbf{q} = \mathbf{p} + \overrightarrow{PQ} = \begin{pmatrix} 0 \\ 15 \\ 3 \end{pmatrix}$ or Q(0,15,3)	
			•2 sets up vector equation for $\mathbf{r}$	•2 $\mathbf{r} = \mathbf{q} + \overrightarrow{QR} = \begin{pmatrix} 0 \\ 15 \\ 3 \end{pmatrix} + \begin{pmatrix} 3 \\ 6 \\ -3 \end{pmatrix}$	
			•3 find coordinate of R or component vector $\mathbf{r}$	•3 $\mathbf{r} = \begin{pmatrix} 3 \\ 21 \\ 0 \end{pmatrix}$ or R(3,21,0)	
			•4 sets up vector equation for $\mathbf{s}$	•4 $\mathbf{s} = \mathbf{r} + \overrightarrow{RS} = \begin{pmatrix} 3 \\ 21 \\ 0 \end{pmatrix} + \begin{pmatrix} 2 \\ 4 \\ -2 \end{pmatrix}$	
			•5 find coordinate of S	•5 S (5,25,−2)	

Question			Generic Scheme. Give one mark for each •	Illustrative Scheme	Max mark
4.			Ans: $-4<p<12$ •1 know discriminant <0 •2 simplify •3 factorise LHS •4 correct range	 •1 $b^2-4ac<0$ and $a=2,\ b=p,\ c=p+6$ stated or implied by •2 •2 $p^2-8p-48<0$ •3 $(p-12)(p+4)<0$ •4 $-4<p<12$	4
5.	(a)		Ans: $m_{l_2}=-\sqrt{3}$ •1 rearranging equation to calculate gradient of line l_1 •2 calculating gradient of l_2	 •1 $y=\frac{1}{\sqrt{3}}x \quad m=\frac{1}{\sqrt{3}}$ •2 $m_{l_2}=-\sqrt{3}$	2
	(b)		Ans: $\theta=\frac{2\pi}{3}$ or 120° •3 using $m=\tan\theta$ •4 calculating angle	 •3 $\tan\theta=-\sqrt{3}$ •4 $\theta=\frac{2\pi}{3}$ or 120°	2
6.	(a) (b)		Ans: $\frac{1+\sqrt{3}}{2\sqrt{2}}$ or $\frac{\sqrt{2}+\sqrt{6}}{4}$ •1 correct expansion •2 any expression equivalent to $\sin 105^\circ$ •3 correct exact value equivalents •4 correct answer	 •1 $\sin x^\circ\cos 60^\circ+\cos x^\circ\sin 60^\circ$ •2 $\sin(45+60)^\circ$ or equivalent •3 $\frac{1}{\sqrt{2}}\times\frac{1}{2}+\frac{1}{\sqrt{2}}\times\frac{\sqrt{3}}{2}$ •4 $\frac{1+\sqrt{3}}{2\sqrt{2}}$ or $\frac{\sqrt{2}+\sqrt{6}}{4}$	4
7.	(a)		•1 know to use $x=-1$ •2 complete synthetic division •3 recognition of zero remainder	•1 -1 | 1 0 −13 −12; −1 1 12; 1 −1 −12 0 •2 -1 | 1 0 −13 −12; −1 1 12; 1 −1 −12 0 •3 $(x+1)$ is a factor as remainder is zero	3
	(b)		Ans: $(x+1)(x+3)(x-4)$ •4 identify quotient •5 factorised fully	 •4 x^2-x-12 •5 $(x+1)(x+3)(x-4)$	2
Notes			Alternative methods of showing $(x+1)$ is a factor, such as long division, inspection and evaluating are perfectly acceptable.		

Question			Generic Scheme. Give one mark for each •	Illustrative Scheme	Max mark
8.	(a)		Ans: $h(x) = 2x^2 - 8x + 5$		3
			•1 correct substitution	•1 $h(x) = 8\left(1 - \frac{1}{2}x\right)^2 - 3$	
			•2 squaring	•2 $1 - x + \frac{1}{4}x^2$	
			•3 expanding and simplifying	•3 $h(x) = 2x^2 - 8x + 5$	
	(b)		Ans: $2(x-2)^2 - 3$		3
			•4 identify common factor	•4 $2(x^2 - 4x$... stated or implied by •3	
			•5 complete the square	•5 $2(x^2 - 2)^2$...	
			•6 process for q	•6 $2(x^2 - 2)^2 - 3$	
Notes			Values for p and q must be consistent with the value for a.		
	(c)		Ans: (2, −3)		1
			•7 state turning point	•7 (2, −3)	
	(d)		Ans:		2
			•8 correct shape	•8 parabola with minimum turning point labelled (positioned consistently with answer to (b))	
			•9 annotation, including y-axis intercept	•9 (0,8)	
9.	(a)		Ans: $y - 10 = -3(x+1)$		1
			•1 finding equation of line	•1 $y - 10 = -3(x+1)$ or equivalent	
	(b)		Ans: B(3,−2)		5
			•2 use of simultaneous equations	•2 $y = -3x + 7$ and $3y = x + 11$	
			•3 solving to find one coordinate of midpoint	•3 either $x = 1$ or $y = 4$	
			•4 finding remaining coordinate of midpoint	•4 M (1, 4)	
			•5 using midpoint formula or ‘stepping out’	•5 either $x = 3$ or $y = -2$	
			•6 finding coordinates of B	•6 B(3, −2)	

Question			Generic Scheme. Give one mark for each •	Illustrative Scheme	Max mark
10.			Ans: $\frac{3\sqrt{3}}{2}$		3
			•1 start to differentiate	•1 $3 \times 4 \sin^2 x$	
			•2 complete differentiation	•2 $\times \cos x$	
			•3 evaluate $f'\left(\frac{5\pi}{6}\right)$	•3 $12\left(\frac{1}{2}\right)^2 \times \frac{-\sqrt{3}}{2} = 12 \times \frac{1}{4} \times \frac{-\sqrt{3}}{2} = \frac{-3\sqrt{3}}{2}$	
11.	(a)		•1 knows derived function represents gradient and that the minimum value of $f'(x)$ is zero	•1 $m = f'(x) \geq 0$ stated explicitly	1
	(b)		•2 interprets information correctly	•2 stationary point plotted in fourth quadrant	2
			•3 completes sketch	•3 point of inflexion on an increasing graph	
12.	(a)		Ans: $\frac{1}{50}$ sec or 0·02 sec		2
			•1 knows how to find period	•1 $T = \frac{2\pi}{100\pi}$	
			•2 correct answer	•2 $\frac{1}{50}$ or 0·02	

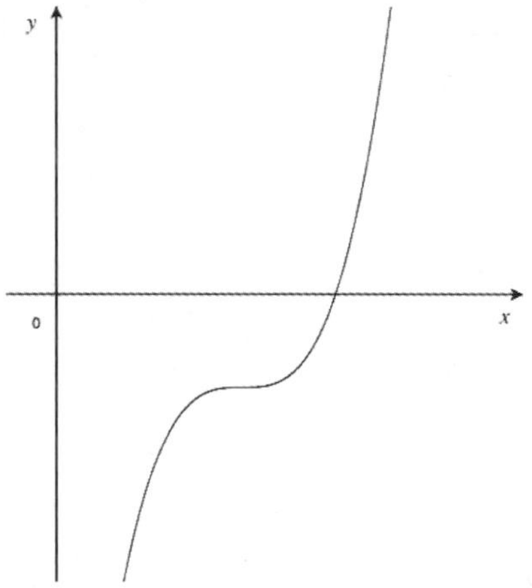

Question			Generic Scheme. Give one mark for each •	Illustrative Scheme	Max mark
	(b)		Ans: $\frac{7}{600}$, $\frac{11}{600}$, and $\frac{19}{600}$ sec		6
			•[1] equating function with –60	•[1] $120\sin 100\pi t = -60$	
			•[2] rearranging	•[2] $\sin 100\pi t = -\frac{1}{2}$	
			•[3] solve equation for $100\pi t$	•[3] $100\pi t = \frac{7\pi}{6}$ and $\frac{11\pi}{6}$	
			•[4] process solutions for t	•[4] $t = \frac{7}{600}$ and $\frac{11}{600}$	
			•[5] knowing to use period or demonstrating another solution from the third quadrant	•[5] $T = \frac{1}{50}$ or $100\pi t = 3\pi + \frac{\pi}{6}$	
			•[6] third value for t	•[6] $\frac{19}{600}$	

Paper 2

Question			Generic Scheme. Give one mark for each •	Illustrative Scheme	Max mark
1.	(a)		•1 find coordinates of E •2 find coordinates of F	•1 E(45, 45, 40) •2 F(60, 40, 0)	2
	(b)		Ans: −1750 •3 find $\overrightarrow{ED}$ and $\overrightarrow{EF}$ •4 correct calculation of scalar product	•3 $\overrightarrow{ED}=\begin{pmatrix}-15\\-15\\40\end{pmatrix}$, $\overrightarrow{EF}=\begin{pmatrix}15\\-5\\-40\end{pmatrix}$ •4 $\overrightarrow{ED}.\overrightarrow{EF}=-225+75-1600=-1750$	2
	(c)		Ans: 154° •5 know how to find angle DEF using formula •6 find $\lvert\overrightarrow{ED}\rvert$ •7 find $\lvert\overrightarrow{EF}\rvert$ •8 calculates angle DEF	•5 $\cos DEF=\frac{\overrightarrow{ED}.\overrightarrow{EF}}{\lvert\overrightarrow{ED}\rvert\lvert\overrightarrow{EF}\rvert}$ or equivalent •6 $\lvert\overrightarrow{ED}\rvert=\sqrt{2050}$ •7 $\lvert\overrightarrow{EF}\rvert=\sqrt{1850}$ •8 $\cos DEF=\frac{-1750}{\sqrt{2050}\sqrt{1850}}$ $DEF=153\cdot977...=154°$	4
2.	(a)		Ans: $a=0\cdot96$, $b=580$ •1 set up one equation •2 set up second equation •3 solve for one variable •4 solve for second variable	•1 $2500=2000a+b$ •2 $2980=2500a+b$ •3 $480=500a$ or $12500=10000a+5b$ $a=\frac{480}{500}$ $11920=10000a+4b$ $580=b$ $a=0\cdot96$ •4 $b=2500-2000\,(0\cdot96)$ $b=2500-1920$ $b=580$ or $2000a=2500-580$ $a=\frac{1920}{2000}$ $a=0\cdot96$	4

Question			Generic Scheme. Give one mark for each •	Illustrative Scheme	Max mark
	(b)		Ans: Yes. Stabilises at 14500		3
			•5 knows how to find the limit	•5 $u_{n+1} = 0\cdot 96u_n + 580, \quad -1 < a < 1$ $L = \dfrac{b}{1-a}$ $L = \dfrac{580}{1-0\cdot 96}$	
			•6 calculate limit	•6 $L = 14500$	
			•7 conclusion	•7 yes, conservation measures will end, since the predicted population stabilises at 14500 and 14500 > 13000	
3.	(a)		Ans: $p=1, q=4$		1
			•1 state values of p and q	•1 $p=1, q=4$	
	(b)		Ans: $y = 9(x-1)$		4
			•2 expand brackets	•2 $f(x) = x^4 - 9x^3 + 24x^2 - 16x$	
			•3 differentiate	•3 $f'(x) = 4x^3 - 27x^2 + 48x - 16$	
			•4 calculate gradient of tangent	•4 $f'(1) = 4 - 27 + 48 - 16 = 9$	
			•5 substitutes gradient and (1,0) into equation of line	•5 $y = 9(x-1)$	
4.	(a)		Ans: $y = \log_4 x + \dfrac{1}{2}$		2
			•1 using law of logarithms	•1 $\log_4 2x = \log_4 2 + \log_4 x$	
			•2 evaluating $\log_4 2$	•2 $\log_4 2 = \dfrac{1}{2}$	
	(b)		Ans: Graph of $y = \log_4 x$ moved up by $\dfrac{1}{2}$ or graph of $y = \log_4 x$ compressed horizontally by a factor of 2.		1
			•3 valid description of relationship	•3 valid description – see answer	
	(c)		Ans: $x = \dfrac{1}{2}$		2
			•4 setting y = 0	•4 $\log_4 2x = 0$	
			•5 solving for x	•5 $x = \dfrac{1}{2}$	

Question			Generic Scheme. Give one mark for each •	Illustrative Scheme	Max mark
	(d)		Ans: 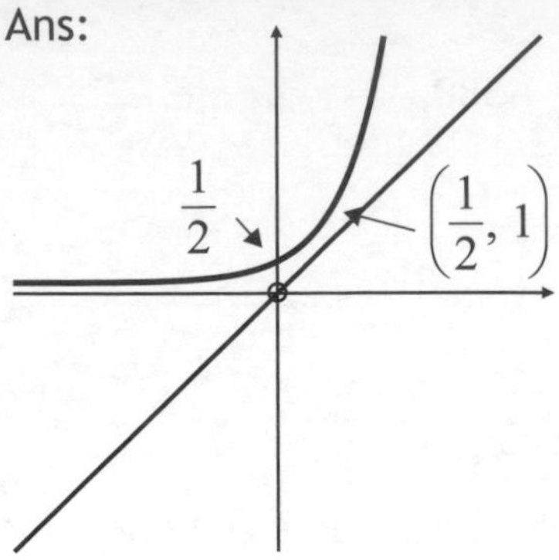 $\frac{1}{2}$; $\left(\frac{1}{2}, 1\right)$		3
			•6 reflecting $y = \log_4 2x$ in the line $y = x$	•6 reflect in $y = x$	
			•7 correct shape	•7 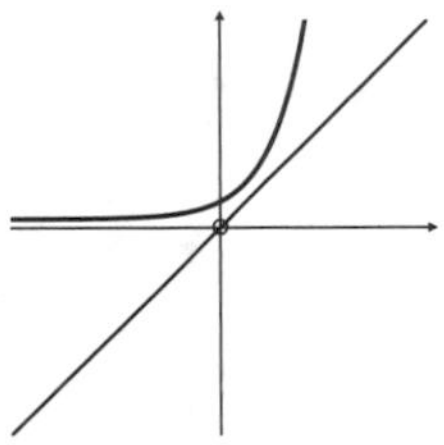	
			•8 annotating (2 points) **(or other valid method)**	•8 $\left(0, \frac{1}{2}\right)$ and $\left(\frac{1}{2}, 1\right)$	
5.	(a)		Ans: $y-1=-2(x-3)$		4
			•1 calculate midpoint of AB	•1 (3, 1)	
			•2 calculate gradient of AB	•2 $\frac{1}{2}$	
			•3 state gradient of perpendicular bisector	•3 -2	
			•4 substitute into equation of line	•4 $y-1=-2(x-3)$	
	(b)		Ans: $(x-2)^2+(y-3)^2=25$		4
			•5 knowing and using $y = 3$	•5 $3 = -2x + 7$	
			•6 solving for x	•6 $x = 2$	
			•7 identifying the radius	•7 $r = 5$	
			•8 obtain circle equation	•8 $(x-2)^2+(y-3)^2=25$	

Question			Generic Scheme. Give one mark for each •	Illustrative Scheme	Max mark
6.			Ans: $x = -3$		5
			•1 use perpendicular property	•1 If $\overrightarrow{CD}$ is perpendicular to $\overrightarrow{AB}$ then $\overrightarrow{CD}.\overrightarrow{AB} = 0$	
			•2 find $\overrightarrow{CD}$	•2 $\begin{pmatrix} x-4 \\ -3 \\ -1 \end{pmatrix}$	
			•3 find $\overrightarrow{AB}$	•3 $\begin{pmatrix} 5 \\ -10 \\ -5 \end{pmatrix}$	
			•4 correct substitution into scalar product formula	•4 $5(x-4)+(-10)(-3)+(-5)(-1)=0$	
			•5 calculates value of x	•5 $x = -3$	
7.			Ans: A False and B True		6
			•1 valid reason for statement A	•1 $P(0) = 30e^{-2} = 4·06$	
			•2 selecting true or false for statement A with valid reason	•2 false, since $P(0) \neq 30$ (do not award without valid reason)	
			•3 setting $P(t) = 15$	•3 $15 = 30e^{t-2}$	
			•4 taking log to base e	•4 $\ln e^{t-2} = \ln 0·5$	
			•5 completing valid reason	•5 $t - 2 = \ln 0·5$ $t = \ln 0·5 + 2$ (1·3)	
			•6 selecting true or false for statement B with valid reason	•6 true, since t = 1·3 to one decimal place and there is only one solution (do not award without valid reason)	
Notes			Substituting $t = 1·3$ into $P(t) = 30e^{t-2}$ is not sufficient to show that statement B is true, since it does not prove that $t = 13$ is the <u>only</u> solution.		
8.	(a)		•1 know to equate volume to 100	•1 $V = \pi r^2 h = 100$	3
			•2 obtain an expression for h	•2 $h = \frac{100}{\pi r^2}$	
			•3 complete area evaluation	•3 $A = \pi r^2 + 2\pi r^2 + 2\pi r \times \frac{100}{\pi r^2}$	

Question			Generic Scheme. Give one mark for each •	Illustrative Scheme	Max mark
	(b)		Ans: r = 2·20 m		6
			•4 know to and start to differentiate	•4 $A'(r) = 6\pi r...$	
			•5 complete differentiation	•5 $A'(r) = 6\pi r - \frac{200}{r^2}$	
			•6 set derivative to zero	•6 $6\pi r - \frac{200}{r^2} = 0$	
			•7 obtain r	•7 $r = 2{\cdot}20$ metres	
			•8 justify nature of stationary point	•8 $A''(r) = 6\pi + \frac{400}{r^3} \Rightarrow A''(2{\cdot}1974...) = 56{\cdot}5...$	
			•9 interpret result	•9 minimum (when $r = 2{\cdot}20$ m)	
Notes			Candidates may use a nature table at •8 to justify a minimum turning point when r = 2·1974...		
9.			Ans: $\frac{5}{6}$		6
			•1 knowing to use integration	•1 $\int \sin\left(\frac{3}{4}x - \frac{3}{2}\pi\right)dx - \int \cos(2x)dx$	
			•2 using correct limits	•2 $\int_0^{\frac{3}{2}\pi} \sin\left(\frac{3}{4}x - \frac{3}{2}\pi\right)dx - \int_0^{\frac{x}{4}} \cos(2x)dx$	
			•3 integrating correctly	•3 $\left[-\frac{4}{3}\cos(\frac{3}{4}x - \frac{3}{2}\pi)\right]......$	
			•4 integrating correctly	•4 $-[\frac{1}{2}\sin(2x)]$	
			•5 substituting limits correctly	•5 See * below	
			•6 evaluating correctly	•6 $(\frac{4}{3} - 0) - (\frac{1}{2} - 0) = \frac{5}{6}$	
* $\left(\left[-\frac{4}{3}\cos\left(\frac{3}{4} \times \frac{3}{2}\pi - \frac{3}{2}\pi\right)\right] - \left[-\frac{4}{3}\cos\left(0 - \frac{3}{2}\pi\right)\right]\right) - \left(\left[\frac{1}{2}\sin\left(2 \times \frac{1}{4}\pi\right)\right]\right) - \left[\frac{1}{2}\sin\left(2 \times 0\right)\right]\right)$					
10.	(a)		Ans: $k = 2, \alpha = \frac{\pi}{6}$ or equivalent		4
			•1 knows to set wave function equal to addition of individual waves	•1 $\sin t + \sqrt{3}\cos t = k\cos(t - \alpha)$ or equivalent	
			•2 knows to expand	•2 $k\cos\alpha\cos t + k\sin\alpha\sin x$ or equivalent	
			•3 knows to compare coefficients	•3 $k\sin\alpha = 1, \quad k\cos\alpha = \sqrt{3}$ or equivalent	
			•4 interpret comparison	•4 $k = 2, \quad \alpha = \frac{\pi}{6}$ or equivalent	

Question			Generic Scheme. Give one mark for each •	Illustrative Scheme	Max mark
	(b)		Ans: 5·9		4
			•5 equates wave function with y-coordinate of P	•5 $2\cos\left(t-\frac{\pi}{6}\right)=1\cdot2$ or equivalent	
			•6 rearranges correctly	•6 $\cos\left(t-\frac{\pi}{6}\right)=0\cdot6$ or equivalent	
			•7 solve equation for $t-\frac{\pi}{6}$	•7 $t-\frac{\pi}{6}=0\cdot927...$ & $5\cdot355...$	
			•8 find t-coordinate of P by interpreting diagram	$1\cdot45...$ & •8 $5\cdot879...$	

HIGHER FOR CfE MATHEMATICS MODEL PAPER

Paper 1 (Non-Calculator)

Question			Generic Scheme. Give one mark for each •	Illustrative Scheme	Max mark
1.			Ans: $2x - y = 5$		3
			•1 know to find m_{AB}	•1 $m_{AB} = \frac{7-3}{4-2}$	
			•2 use $m_{DC} = m_{AB}$	•2 $m_{DC} = 2$	
			•3 find equation of line	•3 $y - 11 = 2(x - 8)$	
2.			Ans: Q (3,1,–2)		3
			•1 interpret ratio	•1 e.g. $\overrightarrow{PQ} = 2\overrightarrow{QR}$	
			•2 start evaluation of components	•2 $3\mathbf{q} = 2\begin{pmatrix}5\\2\\-3\end{pmatrix} + \begin{pmatrix}-1\\-1\\0\end{pmatrix}$	
			•3 complete evaluation of coordinates	•3 Q (3,1,–2)	
3.			Ans: $k = 24$		4
			•1 know discriminant = 0	•1 $b^2 - 4ac = 0$ and $a = k$, $b = k$, $c = 6$ stated or implied by •2	
			•2 simplify	•2 $k^2 - 24k = 0$	
			•3 factorise LHS	•3 $k(k - 24) = 0$	
			•4 solve for given domain (including rejection of $k = 0$)	•4 $k = 24$	
4.			Ans: **u.v**=0 ⇒ **u** and **v** are perpendicular		2
			•1 use scalar product	•1 $3 \times 2 + 2 \times (-3) + 0 \times 4$	
			•2 communicate solution	•2 **u.v**=0 ⇒ **u** and **v** are perpendicular	
5.			Ans: $2x^2 + \frac{1}{x} + c$		4
			•1 preparation for integration	•1 $4x - x^{-2}$	
			•2 correct integration of first term	•2 $4\frac{x^2}{2} - \ldots\ldots$	
			•3 correct integration of second term	•3 $\ldots\ldots - \frac{x^{-1}}{-1}$	
			•4 includes constant of integration	•4 $2x^2 + x^{-1} + c$	
6.	(a)		Ans: $\frac{x-3}{2}$		2
			•1 change subject to x	•1 $y = 2x + 3 \Rightarrow x = \frac{y-3}{2}$	
			•2 express function in terms of x	•2 $\frac{x-3}{2}$	

Question			Generic Scheme. Give one mark for each •	Illustrative Scheme	Max mark
	(b)	(i)	Ans: $\frac{1}{2x-1}$		2
			•1 start composite process	•1 e.g. $f(2x + 3)$	
			•2 find $h(x)$ in simplest form	•2 $\frac{1}{2x-1}$	
		(ii)	Ans: $x \neq \frac{1}{2}$		1
			•1 state restriction	•1 $x \neq \frac{1}{2}$	
7.			Ans: $g^2 + f^2 - c = -\frac{7}{4}$, so equation does not represent a circle since $g^2 + f^2 - c < 0$		2
			•1 use $g^2 + f^2 - c$	•1 $g^2 + f^2 - c = 1^2 + \left(\frac{3}{2}\right)^2 - 5$	
			•2 communicate solution	•2 $g^2 + f^2 - c = -\frac{7}{4}$, so equation does not represent a circle since $g^2 + f^2 - c < 0$.	
8.	(a)		Ans: 9·8m/s		3
			•1 know to differentiate	•1 $\frac{dh}{dt} = \ldots\ldots$	
			•2 differentiate correctly	•2 $\frac{dh}{dt} = 19{\cdot}6 - 9{\cdot}8t$	
			•3 find speed after 1 second	•3 9·8m/s	
	(b)		Ans: 2 seconds		2
			•1 know how to find stationary point	•1 e.g. $19{\cdot}6 - 9{\cdot}8t = 0$	
			•2 communicate solution	•2 2 seconds	
9.	(a)		Ans: $8 - (x + 1)^2$		2
			•1 complete the square	•1 $\ldots -(x + 1)^2$	
			•2 process for a	•2 $8 -(x + 1)^2$	
	(b)		Ans: 8; maximum value occurs when $(x + 1)^2 = 0$.		2
			•1 state maximum value	•1 8	
			•2 justification	•2 e.g. maximum value occurs when $(x + 1)^2 = 0$.	
10.			Ans: $a = -2$, $b = 5$		3
			•1 use fact that graph of $y = \log_b x$ passes through (1,0)	•1 graph of $y = \log_b (x + a)$ is graph of $y = \log_b x$ moved 2 units right	
			•2 state value of a	•2 −2	
			•3 state value of b	•3 5	

Question			Generic Scheme. Give one mark for each •	Illustrative Scheme	Max mark
11.			Ans:		3
			•[1] identify roots	•[1] 0 and c only	
			•[2] interpret point of inflection	•[2] turning point at (0,0)	
			•[3] complete cubic curve	•[3] correct shape	
12.	(a)		Ans: (−1, 4) maximum turning point; (1, 0) minimum turning point		6
			•[1] differentiate correctly	•[1] $f'(x) = 3x^2 - 3$	
			•[2] set derivative to zero	•[2] $3x^2 - 3 = 0$	
			•[3] solve for x	•[3] $x = -1$, $x = 1$	
			•[4] evaluate y-coordinates	•[4] $y = 4$, $y = 0$	
			•[5] justify nature of stationary points	•[5] x ..− 1.. ..1.. $f'(x)$ + 0 − −0+	
			•[6] interpret result	•[6] (−1, 4) max. turning point; (1, 0) min. turning point	
	(b)	(i)	Ans: **proof**		3
			•[1] know to use $x = 1$	•[1] 1 | 1 0 −3 2	
			•[2] complete synthetic division	•[2] 1 | 1 0 −3 2 0 1 1 2 1 1 −2 0	
			•[3] recognition of zero remainder	•[3] $(x - 1)$ is a factor as remainder is zero	
		(ii)	Ans: $(x-1)(x-1)(x+2)$		2
			•[1] identify quotient	•[1] $x^2 + x + 2$	
			•[2] factorise fully	•[2] $(x-1)(x-1)(x+2)$	

Question			Generic Scheme. Give one mark for each •	Illustrative Scheme	Max mark
	(c)		Ans: x-axis (1,0) and (−2,0); y-axis (0,2). 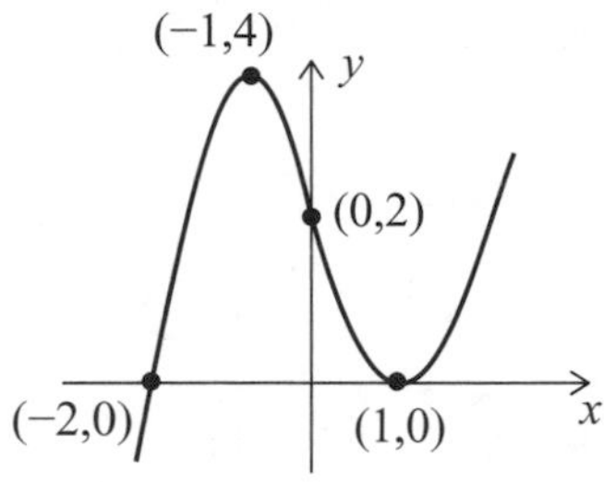		4
			(−1,4) y (0,2) (−2,0) (1,0) x		
			•[1] state y-intercepts	•[1] (0,2)	
			•[2] state x-intercepts	•[2] (1,0) and (−2,0)	
			•[3] sketch of correct cubic curve with stationary points shown and labelled	•[3] sketch of correct cubic curve with max. t.p.(−1,4) and min. t.p. (1,0) shown and labelled	
			•[4] intercepts with axes shown and labelled on sketch	•[4] intercepts (0,2),(1,0) and (−2,0) shown and labelled on sketch	
13.			Ans: $-\frac{3}{5}$		7
			•[1] express DEA in terms of x°	•[1] $(2x + 90)^\circ$	
			•[2] expand $\cos (2x + 90)^\circ$	•[2] $\cos 2x^\circ \cos 90^\circ - \sin 2x^\circ \sin 90^\circ$	
			•[3] simplify	•[3] $-\sin 2x^\circ$	
			•[4] expand $\sin 2x^\circ$	•[4] $-2 \sin x^\circ \cos x^\circ$	
			•[5] find exact length of CE	•[5] $\sqrt{10}$	
			•[6] substitute into $-2 \sin x^\circ \cos x^\circ$	•[6] $-2 \times \frac{1}{\sqrt{10}} \times \frac{3}{\sqrt{10}}$	
			•[7] find exact value of cos DEA	•[7] $-\frac{3}{5}$	

Paper 2

Question			Generic Scheme. Give one mark for each •	Illustrative Scheme	Max mark
1.	(a)		Ans: $y = 2x - 5$		4
			•1 find midpoint of BC	•1 (1,−3)	
			•2 find gradient of BC	•2 $m_{BC} = -\frac{1}{2}$	
			•3 find perpendicular gradient	•3 $m_{perp} = 2$	
			•4 show steps leading to equation of perpendicular bisector in given form	•4 $y-(-3) = 2(x-1)$ $y+3 = 2x-2$ $y = 2x-5$	
	(b)		Ans: $3x + y = 10$		3
			•1 find midpoint of AB	•1 (2,4)	
			•2 find gradient of median	•2 $m_{median} = -3$	
			•3 find equation of median	•3 $y-(-5) = -3(x-5)$	
	(c)		Ans: (3,1)		3
			•1 use valid approach	•1 e.g. $2x - 5 = -3x + 10$	
			•2 solve for one variable	•2 $x = 3$ or $y = 1$	
			•3 find coordinates of point of intersection	•3 (3,1)	
2.	(a)		Ans: **B**(4,4,0)		1
			•1 state coordinates of B	•1 (4,4,0)	
	(b)		Ans: $\overrightarrow{DB} = \begin{pmatrix} 2 \\ 2 \\ -6 \end{pmatrix}$, $\overrightarrow{DM} = \begin{pmatrix} 0 \\ -2 \\ -6 \end{pmatrix}$		3
			•1 state components of $\overrightarrow{DB}$	•1 $\begin{pmatrix} 2 \\ 2 \\ -6 \end{pmatrix}$	
			•2 find coordinates of M	•2 (2,0,0) stated or implied by •3	
			•3 state components of $\overrightarrow{DM}$	•3 $\begin{pmatrix} 0 \\ -2 \\ -6 \end{pmatrix}$	
	(c)		Ans: 40·3° or 0·703 rads		5
			•1 know to use scalar product	•1 $\cos BDM = \frac{\overrightarrow{DB}.\overrightarrow{DM}}{\lvert\overrightarrow{DB}\rvert\lvert\overrightarrow{DM}\rvert}$	
			•2 find scalar product	•2 $\overrightarrow{DB}.\overrightarrow{DM} = 32$	
			•3 find magnitude of a vector	•3 $\lvert\overrightarrow{DB}\rvert = \sqrt{44}$	
			•4 find magnitude of a vector	•4 $\lvert\overrightarrow{DM}\rvert = \sqrt{40}$	
			•5 evaluate angle BDM	•5 40·3° or 0·703 rads	

Question			Generic Scheme. Give one mark for each •	Illustrative Scheme	Max mark
3.	(a)		Ans: 2·5 metres		3
			•1 use recurrence relation	•1 $u_{n+1} = 0{\cdot}8u_n + 0{\cdot}5$	
			•2 know how to find limit	•2 $L = \frac{0{\cdot}5}{1-0{\cdot}8}$	
			•3 calculate limit	•3 2·5 metres	
	(b)		Ans: 25%		3
			•1 set up equation involving limit	•1 $\frac{0{\cdot}5}{1-b} = 2$	
			•2 solve equation	•2 $b = 0{\cdot}75$	
			•3 state percentage	•3 25%	
4.			Ans: $21\frac{1}{12}$ or $\frac{253}{12}$ or 21·1		10
			•1 know to integrate	•1 $\int$.... or attempt integration	
			•2 know to deal with areas on each side of y-axis	•2 evidence of attempting to interpret the diagram to left of y-axis separately from diagram to right	
			•3 interpret limits of one area	•3 e.g. $\int_{-2}^{0}$	
			•4 use "upper – lower"	•4 $(x^3 - x^2 - 4x + 4) - (2x + 4)$	
			•5 integrate	•5 $\frac{1}{4}x^4 - \frac{1}{3}x^3 - 3x^2$	
			•6 substitute in limits	•6 $-(\frac{1}{4}(-2)^4 - \frac{1}{3}(-2)^3 - 3(-2)^2)$ Evidence for •6 may be implied by •7 but •7 must be consistent with •5	
			•7 evaluate the area on one side	•7 $\frac{16}{3}$	
			•8 interpret integrand with the limits of the other area	•8 $\int_{0}^{3} (2x - 4) - (x^3 - x^2 - 4x + 4)dx$	
			•9 evaluate the area on the other side	•9 $\frac{63}{4}$	
			•10 state total area	•10 $21\frac{1}{12}$ or $\frac{253}{12}$ or 21·1	
5.	(a)		Ans: $k = 13, a = 22{\cdot}6$		4
			•1 use addition formula	•1 $k \cos x° \cos a° - k \sin° x \sin a°$ or $k(\cos x° \cos a° - \sin° \sin a°)$	
			•2 compare coefficients	•2 $k \cos a° = 12$ **and** $k \sin a° = 5$ or $-k \sin a° = -5$	
			•3 process k	•3 13	
			•4 process a	•4 22·6	
	(b)	(i)	Ans: max. = 13, min. = −13		1
			•1 state maximum and minimum	•1 13, −13	

Question			Generic Scheme. Give one mark for each •	Illustrative Scheme	Max mark
		(ii)	Ans: max. at 337·4, min. at 157·4		2
			•[1] find x corresponding to maximum value	•[1] maximum at 337·4	
			•[2] find x corresponding to minimum value	•[2] minimum at 157·4	
6.			Ans: 0·363		4
			•[1] start to integrate	•[1] $-\cos(4x+1)$	
			•[2] complete integration	•[2] $-\frac{1}{4}\cos(4x+1)$	
			•[3] substitute in limits	•[3] $-\frac{1}{4}\cos 9 - (-\frac{1}{4}\cos 1)$	
			•[4] evaluate integral	•[4] 0·363	
7.	(a)	(i)	Ans: proof	**Method 1**	4
			•[1] substitute	•[1] $x^2 + (3-x)^2 + 14x + 4(3-x) - 19 = 0$	
			•[2] express in standard form	•[2] $2x^2 + 4x + 2 = 0$	
			•[3] start proof	•[3] $2(x+1)(x+1) = 0$	
			•[4] complete proof	•[4] equal roots so line is a tangent	
				Method 2	
				•[1] $x^2 + (3-x)^2 + 14x + 4(3-x) - 19 = 0$	
				•[2] $2x^2 + 4x + 2 = 0$	
				•[3] $b^2 - 4ac = 4^2 - 4 \times 2 \times 2$	
				•[4] $b^2 - 4ac = 0$ so line is a tangent	
		(ii)	Ans: $(-1,4)$		1
			•[1] state coordinates of P	•[1] $(-1,4)$	

Question			Generic Scheme. Give one mark for each •	Illustrative Scheme	Max mark
7.	(b)		Ans: $(x-1)^2+(y-6)^2=8$ or $x^2+y^2-2x-12y=29=0$		6
			Method 1	**Method 1**	
			•1 state centre of larger circle	•1 $(-7,-2)$	
			•2 find radius of larger circle	•2 $\sqrt{72}$	
			•3 find radius of smaller circle	•3 $\sqrt{8}$	
			•4 strategy for finding centre	•4 e.g. "stepping out"	
			•5 interpret centre of smaller circle	•5 $(1,6)$	
			•6 state equation	•6 $(x-1)^2+(y-6)^2=8$ or $x^2+y^2-2x-12y=29=0$	
			Method 2	**Method 2**	
			•1 state centre of larger circle	•1 $(-7,-2)$	
			•2 strategy for finding centre	•2 e.g. "stepping out"	
			•3 state centre of smaller circle	•3 $(1,6)$	
			•4 strategy for finding radius	•4 $\sqrt{2^2+2^2}$	
			•5 find radius of smaller circle	•5 $\sqrt{8}$	
			•6 state equation	•6 $(x-1)^2+(y-6)^2=8$ or $x^2+y^2-2x-12y=29=0$	
8.	(a)		Ans: 4433 micrograms		3
			•1 interpret equation	•1 $600 = A_0e^{-0.002\times1000}$	
			•2 process equation	•2 $A_0 = \dfrac{600}{e^{-0.002\times1000}}$	
			•3 process for A_0	•3 $A_0 \approx 4433$ micrograms	
	(b)		Ans: 346·6 years		5
			•1 interpret half-life	•1 $\frac{1}{2}A_0 = A_0e^{-0.002t}$	
			•2 process equation	•2 $e^{-0.002t} = \frac{1}{2}$	
			•3 take log to base e	•3 $\ln e^{-0.002t} = \ln\frac{1}{2}$	
			•4 process log equation	•4 $t = \dfrac{\log_e \frac{1}{2}}{-0{\cdot}002}$	
			•5 process for t	•5 346·6 years	

Question			Generic Scheme. Give one mark for each •	Illustrative Scheme	Max mark
9.			Ans: 2·419, 3·864 and no solution		5
			•¹ use double angle formula	•¹ $2 \times (2\cos^2 x - 1)\ldots\ldots$	
			•² express as quadratic in $\cos x$	•² $4\cos^2 x - 5\cos x - 6 = 0$	
			•³ start to solve	•³ $(4\cos x + 3)(\cos x - 2) = 0$ or $\cos x = \frac{(-5)^2 \pm \sqrt{4 \times 4 \times (-6)}}{2 \times 4}$	
			•⁴ solve for $\cos x$	•⁴ $\cos x = -\frac{3}{4}$ and $\cos x = 2$	
			•⁵ solve for x	•⁵ 2·419, 3·864 and no solution	

HIGHER MATHEMATICS 2015

Paper 1 (Non-Calculator)

Question			Generic Scheme	Illustrative Scheme	Max mark
1.			•[1] equate scalar product to zero •[2] state value of t	•[1] $-24+2t+6=0$ •[2] $t=9$	2
2.			•[1] know to and differentiate •[2] evaluate $\frac{dy}{dx}$ •[3] evaluate y-coordinate •[4] state equation of tangent	•[1] $6x^2$ •[2] 24 •[3] -13 •[4] $y=24x+35$	4
3.			 •[1] know to use $x=-3$ •[2] interpret result and state conclusion •[3] state quadratic factor •[4] factorise completely	**Method 1** •[1] $(-3)^3-3(-3)^2-10(-3)+24$ •[2] $=0 \therefore (x+3)$ is a factor. **Method 2** •[1] $\begin{array}{r\|rrrr} -3 & 1 & -3 & -10 & 24 \\ & & -3 & & \\ \hline & 1 & & & \end{array}$ •[2] $\begin{array}{r\|rrrr} -3 & 1 & -3 & -10 & 24 \\ & & -3 & 18 & -24 \\ \hline & 1 & -6 & 8 & 0 \end{array}$ remainder $=0 \therefore (x+3)$ is a factor. **Method 3** •[1] $x+3\overline{)x^3-3x^2-10x+24}$ with quotient x^2; x^3+3x^2 •[2] $=0 \therefore (x+3)$ is a factor. •[3] x^2-6x+8 stated or implied by •[4] •[4] $(x+3)(x-4)(x-2)$	4
4.			•[1] state the value of p •[2] state the value of q •[3] state the value of r	•[1] $p=3$ •[2] $q=4$ •[3] $r=1$	3

Question			Generic Scheme	Illustrative Scheme	Max mark
5.	(a)		•[1] let $y = 6 - 2x$ and rearrange. •[2] state expression. **Method 2** •[3] equates composite function to x •[1] start to rearrange. •[2] state expression.	•[1] $x = \frac{6-y}{2}$ or $y = \frac{6-x}{2}$ •[2] $g^{-1}(x) = \frac{6-x}{2}$ or $3 - \frac{x}{2}$ or $\frac{x-6}{-2}$ **Method 2** $g(g^{-1}(x)) = x$ this gains •[3] $6 - 2g^{-1}(x) = x$ $g^{-1}(x) = \frac{6-x}{2}$ or $3 - \frac{x}{2}$ or $\frac{x-6}{-2}$	2
	(b)		•[3] state expression	•[3] x	1
6.			•[1] use laws of logs •[2] use laws of logs •[3] evaluate log	•[1] $\log_6 27^{\frac{1}{3}}$ •[2] $\log_6\left(12 \times 27^{\frac{1}{3}}\right)$ •[3] 2	3
7.			•[1] write in differentiable form •[2] differentiate first term •[3] differentiate second term •[4] evaluate derivative at $x = 4$	•[1] $3x^{\frac{3}{2}} - 2x^{-1}$ •[2] $\frac{9}{2}x^{\frac{1}{2}} + \ldots$ •[3] $\ldots + 2x^{-2}$ •[4] $9\frac{1}{8}$	4
8.			•[1] interpret information •[2] express in standard quadratic form •[3] factorise •[4] state range	•[1] $x(x-2) < 15$ •[2] $x^2 - 2x - 15 < 0$ •[3] $(x-5)(x+3) < 0$ •[4] $2 < x < 5$	4
9.			•[1] find gradient of AB •[2] calculate gradient of BC •[3] interpret results and state conclusion	•[1] $m_{AB} = -\sqrt{3}$ •[2] $m_{BC} = -\frac{1}{\sqrt{3}}$ •[3] $m_{AB} \neq m_{BC} \Rightarrow$ points are not collinear. **Method 2** •[1] $m_{AB} = -\sqrt{3}$ •[2] AB makes $120°$ with positive direction of the x-axis. •[3] $120 \neq 150$ so points are not collinear.	3

Question			Generic Scheme	Illustrative Scheme	Max mark
10.	(a)		•1 state value of $\cos 2x$	•1 $\frac{4}{5}$	**1**
	(b)		•2 use double angle formula •3 evaluate $\cos x$	•2 $2\cos^2 x - 1 = \ldots$ •3 $\frac{3}{\sqrt{10}}$	**2**
11.	(a)		•1 state coordinates of centre •2 find gradient of radius •3 state perpendicular gradient •4 determine equation of tangent	•1 $(-8,-2)$ •2 $-\frac{1}{2}$ •3 2 •4 $y = 2x - 1$	**4**
	(b)		**Method 1** •5 arrange equation of tangent in appropriate form and equate y_{tangent} to y_{parabola} •6 rearrange and equate to 0 •7 know to use discriminant and identify a, b, and c •8 simplify and equate to 0 •9 start to solve •10 state value of p **Method 2** •5 arrange equation of tangent in appropriate form and equate y_{tangent} to y_{parabola} •6 find $\frac{dy}{dx}$ for parabola •7 equate to gradient of line and rearrange for p •8 substitute and arrange in standard form •9 factorise and solve for x •10 state value of p	**Method 1** •5 $2x - 1 = -2x^2 + px + 1 - p$ •6 $2x^2 + (2-p)x + p - 2 = 0$ •7 $(2-p)^2 - 4 \times 2 \times (p-2)$ •8 $p^2 - 12p + 20 = 0$ •9 $(p-10)(p-2) = 0$ •10 $p = 10$ **Method 2** •5 $2x - 1 = -2x^2 + px + 1 - p$ •6 $\frac{dy}{dx} = -4x + p$ •7 $2 = -4x + p$ $p = 2 + 4x$ •8 $0 = 2x^2 - 4x$ •9 $0 = 2x(x-2)$ $x = 0,\ x = 4$ •10 $p = 10$	**6**
12.			•1 interpret integral below x-axis •2 evaluate	•1 -1 (accept area below x-axis $= 1$) •2 $-\frac{1}{2}$	**2**

Question			Generic Scheme	Illustrative Scheme	Max mark
13.	(a)		•1 calculate b	•1 5	1
	(b)	(i)	•2 reflecting in the line $y = x$	•2 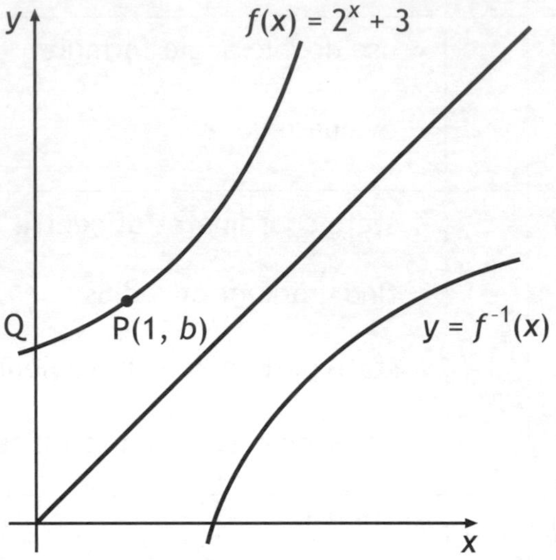	1
	(b)	(ii)	•3 calculate y intercept •4 state coordinates of image of Q •5 state coordinates of image of P	•3 4 •4 (4, 0) see note 2 •5 (5, 1)	3
	(c)		•6 state x coordinate of R •7 state y coordinate of R	•6 $x = 2$ •7 $y = -7$	2
14.			•1 identify length of radius •2 determine value of k	y-axis tangent to circle: •1 $r = 6$; •2 $k = 25$ Circle passes through origin: $r = \sqrt{61}$; $k = 0$	2
15.			•1 know to integrate •2 integrate a term •3 complete integration •4 find constant of integration •5 find value of k •6 state expression for T	•1 $\int$ •2 $\frac{1}{50}t^2 \ldots$ or $\ldots - kt$ •3 $\ldots - kt$ or $\frac{1}{50}t^2 \ldots$ •4 $c = 100$ •5 $k = 2$ •6 $T = \frac{1}{50}t^2 - 2t + 100$	6

Paper 2

Question			Generic Scheme	Illustrative Scheme	Max mark
1.	(a)		•1 calculate gradient of AB •2 use property of perpendicular lines •3 substitute into general equation of a line •4 demonstrate result	•1 $m_{AB} = -3$ •2 $m_{alt} = \frac{1}{3}$ •3 $y - 3 = \frac{1}{3}(x - 13)$ •4 $\ldots \Rightarrow x - 3y = 4$	4
	(b)		•5 calculate midpoint of AC •6 calculate gradient of median •7 determine equation of median	•5 $\text{M}_{\text{AC}} = (4,5)$ •6 $m_{BM} = 2$ •7 $y = 2x - 3$	3
	(c)		•8 calculate x or y coordinate •9 calculate remaining coordinate of the point of intersection	•8 $x = 1$ or $y = -1$ •9 $y = -1$ or $x = 1$	2
2.	(a)		•1 interpret notation •2 state a correct expression	•1 $f((1+x)(3-x)+2)$ stated or implied by •2 •2 $10 + (1+x)(3-x) + 2$ stated or implied by •3	2
	(b)		•3 write $f(g(x))$ in quadratic form **Method 1** •4 identify common factor •5 complete the square **Method 2** •4 expand completed square form and equate coefficients •5 process for q and r and write in required form	•3 $15 + 2x - x^2$ or $-x^2 + 2x + 15$ **Method 1** •4 $-1(x^2 - 2x$ stated or implied by •5 •5 $-1(x-1)^2 + 16$ **Method 2** •4 $px^2 + 2pqx + pq^2 + r$ and $p = -1$, •5 $q = -1$ and $r = 16$ Note if $p = 1$ •5 is not available	3
	(c)		•6 identify critical condition •7 identify critical values	•6 $-1(x-1)^2 + 16 = 0$ or $f((g(x)) = 0$ •7 5 and -3	2

Question			Generic Scheme	Illustrative Scheme	Max mark
3.	(a)		•1 determine the value of the required term	•1 $22\frac{3}{4}$ or $\frac{91}{4}$ or 22·75	1
	(b)		**Method 1** **(Considering both limits)**	**Method 1**	5
			•2 know how to calculate limit	•2 $\frac{32}{1-\frac{1}{3}}$ or $L=\frac{1}{3}L+32$	
			•3 know how to calculate limit	•3 $\frac{13}{1-\frac{3}{4}}$ or $L=\frac{3}{4}L+13$	
			•4 calculate limit	•4 48	
			•5 calculate limit	•5 52	
			•6 interpret limits and state conclusion	•6 $52>50$ $\therefore$ toad will escape	
			Method 2 **(Frog first then numerical for toad)**	**Method 2**	
			•2 know how to calculate limit	•2 $\frac{32}{1-\frac{1}{3}}$ or $L=\frac{1}{3}L+32$	
			•3 calculate limit	•3 48	
			•4 determine the value of the highest term less than 50	•4 49·803...	
			•5 determine the value of the lowest term greater than 50	•5 50·352...	
			•6 interpret information and state conclusion	•6 50·352>50 $\therefore$ toad will escape	
			Method 3 **(Numerical method for toad only)**	**Method 3**	
			•2 continues numerical strategy	•2 numerical strategy	
			•3 exact value	•3 30·0625	
			•4 determine the value of the highest term less than 50	•4 49·803...	
			•5 determine the value of the lowest term greater than 50	•5 50·352...	
			•6 interpret information and state conclusion	•6 50·352>50 $\therefore$ toad will escape	
			Method 4 **(Limit method for toad only)**	**Method 4**	
			•2 & •3 know how to calculate limit	•2 & •3 $\frac{13}{1-\frac{3}{4}}$ or $L=\frac{3}{4}L+13$	
			•4 & •5 calculate limit	•4 & •5 52	
			•6 interpret limit and state conclusion	•6 52>50 $\therefore$ toad will escape	

Question			Generic Scheme	Illustrative Scheme	Max mark
4.	(a)		•1 know to equate $f(x)$ and $g(x)$ •2 solve for x	•1 $\frac{1}{4}x^2-\frac{1}{2}x+3=\frac{1}{4}x^2-\frac{3}{2}x+5$ •2 $x=2$	**2**
	(b)		•3 know to integrate •4 interpret limits •5 use 'upper - lower' •6 integrate •7 substitute limits •8 evaluate area between $f(x)$ and $h(x)$ •9 state total area	•3 $\int$ •4 $\int_0^2$ •5 $\int_0^2\left(\frac{1}{4}x^2-\frac{1}{2}x+3\right)-\left(\frac{3}{8}x^2-\frac{9}{4}x+3\right)dx$ •6 $-\frac{1}{24}x^3+\frac{7}{8}x^2$ accept unsimplified integral •7 $\left(-\frac{1}{24}\times2^3+\frac{7}{8}\times2^2\right)-0$ •8 $\frac{19}{6}$ •9 $\frac{19}{3}$	**7v**
5.	(a)		•1 state centre of C_1 •2 state radius of C_1 •3 calculate distance between centres of C_1 and C_2 •4 calculate radius of C_2	•1 $(-3,-5)$ •2 5 •3 20 •4 15	**4**
	(b)		•5 find ratio in which centre of C_3 divides line joining centres of C_1 and C_2 •6 determine centre of C_3 •7 calculate radius of C_3 •8 state equation of C_3	•5 $3:1$ •6 $(6,7)$ •7 $r=20$ (answer must be consistent with distance between centres) •8 $(x-6)^2+(y-7)^2=400$	**4**
6.	(a)		•1 Expands •2 Evaluate $\mathbf{p}\cdot\mathbf{q}$ •3 Completes evaluation	•1 $\mathbf{p}\cdot\mathbf{q}+\mathbf{p}\cdot\mathbf{r}$ •2 $4\frac{1}{2}$ •3 $\ldots+0=4\frac{1}{2}$	**3**
	(b)		•4 correct expression	•4 $-\mathbf{q}+\mathbf{p}+\mathbf{r}$ or equivalent	**1**
	(c)		•5 correct substitution •6 start evaluation •7 find expression for $\lvert\mathbf{r}\rvert$	•5 $-\mathbf{q}\cdot\mathbf{q}+\mathbf{q}\cdot\mathbf{p}+\mathbf{q}\cdot\mathbf{r}$ •6 $-9+\ldots+3\lvert\mathbf{r}\rvert\cos30^\circ=9\sqrt{3}-\frac{9}{2}$ •7 $\lvert\mathbf{r}\rvert=\frac{3\sqrt{3}}{\cos 30}$	**3**

Question			Generic Scheme	Illustrative Scheme	Max mark
7.	(a)		•1 integrate a term	•1 $\frac{3}{2}\sin 2x$ OR x	2
			•2 complete integration with constant	•2 $x + c$ ⁞ $\frac{3}{2}\sin 2x + c$	
	(b)		•3 substitute for cos $2x$	•3 $3(\cos^2 x - \sin^2 x)\ldots$ or $\ldots(\sin^2 x + \cos^2 x)$	2
			•4 substitute for 1 and complete	•4 $\ldots(\sin^2 x + \cos^2 x) = 4\cos^2 x - 2\sin^2 x$	
	(c)		•5 interpret link	•5 $-\frac{1}{2}\int\ldots$	2
			•6 state result	•6 $-\frac{3}{4}\sin 2x - \frac{1}{2}x + c$	
8.	(a)	(i)	•1 calculate T when $x = 20$	•1 10·4 or 104	1
		(ii)	•2 calculate T when $x = 0$	•2 11 or 110	1
	(b)		•3 write function in differential form	•3 $5(36 + x^2)^{\frac{1}{2}} + \ldots$	8
			•4 start differentiation of first term	•4 $5 \times \frac{1}{2}(\quad)^{-\frac{1}{2}}\ldots$	
			•5 complete differentiation of first term	•5 $\ldots\ldots\ldots\ldots \times 2x\ldots\ldots$	
			•6 complete differentiation and set candidate's derivative = 0	•6 $5x(36 + x^2)^{-\frac{1}{2}} - 4 = 0$	
			•7 start to solve	•7 $5x = 4(36 + x^2)^{\frac{1}{2}}$ or $\frac{5x}{(36 + x^2)^{\frac{1}{2}}} = 4$	
			•8 know to square both sides	•8 $25x^2 = 16(36 + x^2)$ or $\frac{25x^2}{(36 + x^2)} = 16$	
			•9 find value of x	•9 $x = 8$	
			•10 calculate minimum time	•10 T = 9·8 or 98 no units required	

Question			Generic Scheme	Illustrative Scheme	Max mark
9.			•1 use compound angle formula	•1 $k\sin 1{\cdot}5t\cos a - k\cos 1{\cdot}5t\sin a$	8
			•2 compare coefficients	•2 $k\cos a = 36$, $k\sin a = 15$ **stated explicitly**	
			•3 process for k	•3 $k = 39$	
			•4 process for a	•4 $a = 0{\cdot}39479\ldots$rad or $22{\cdot}6^\circ$	
			•5 equates expression for h to 100	•5 $39\sin\left(1{\cdot}5t - 0{\cdot}39479\ldots\right) + 65 = 100$	
			•6 write in standard format and attempt to solve	•6 $\sin\left(1{\cdot}5t - 0{\cdot}39479\ldots\right) = \frac{35}{39}$ $\Rightarrow 1{\cdot}5t - 0{\cdot}39479\ldots = \sin^{-1}\left(\frac{35}{39}\right)$ •7 ⋮ •8	
			•7 solve equation for $1{\cdot}5t$	•7 $1{\cdot}5t = 1{\cdot}508$ and $2{\cdot}422$	
			•8 process solutions for t	•8 $t = 1{\cdot}006$ and $1{\cdot}615$	

HIGHER MATHEMATICS 2016

Paper 1 (Non-Calculator)

Question			Generic Scheme	Illustrative Scheme	Max mark
1.			•1 find the gradient •2 state equation	•1 -4 •2 $y + 4x = -5$	2
2.			•1 write in differentiable form •2 differentiate first term •3 differentiate second term	•1 ... $+ 8x^{\frac{1}{2}}$ stated or implied by •3 •2 $36x^2$ •3 $4x^{-\frac{1}{2}}$	3
3.	(a)		•1 interpret recurrence relation and calculate u_4	•1 $u_4 = 12$	1
	(b)		•2 communicate condition for limit to exist	•2 A limit exists as the recurrence relation is linear and $-1 < \frac{1}{3} < 1$	1
	(c)		•3 know how to calculate limit •4 calculate limit	•3 $\frac{10}{1-\frac{1}{3}}$ or $L = \frac{1}{3}L + 10$ •4 15	2
4.			•1 find the centre •2 calculate the radius •3 state equation of circle	•1 $(-3, 4)$ stated or implied by •3 •2 $\sqrt{17}$ •3 $(x + 3)^2 + (y - 4)^2 = 17$ or equivalent	3
5.			•1 start to integrate •2 complete integration	•1 ... $\times \sin(4x + 1)$ •2 $2\sin(4x + 1) + c$	2
6.	(a)		•1 equate composite function to x •2 write $f(f^{-1}(x))$ in terms of $f^{-1}(x)$ •3 state inverse function	**Method 1** •1 $f(f^{-1}(x)) = x$ •2 $3f^{-1}(x) + 5 = x$ •3 $f^{-1}(x) = \frac{x-5}{3}$	3
			•1 write as $y = 3x + 5$ start to rearrange •2 complete rearrangement •3 state inverse function	**Method 2** •1 $y - 5 = 3x$ •2 $x = \frac{y-5}{3}$ •3 $f^{-1}(x) = \frac{x-5}{3}$	3

Question			Generic Scheme	Illustrative Scheme	Max mark
				Method 3	3
			•1 interchange variables	•1 $x = 3y + 5$	
			•2 complete rearrangement	•2 $\frac{x-5}{3} = y$	
			•3 state inverse function	•3 $f^{-1}(x) = \frac{x-5}{3}$	
	(b)		•4 correct value	•4 2	1
7.	(a)		•1 identify pathway	•1 $\overrightarrow{FG} + \overrightarrow{GH}$	2
			•2 state $\overrightarrow{FH}$	•2 **i** + 3**j** – 4**k**	
	(b)		•3 identify pathway	•3 $\overrightarrow{FH} + \overrightarrow{HE}$ or equivalent	2
			•4 $\overrightarrow{FE}$	•4 –**i** – 5**k**	
8.			•1 substitute for y	•1 $x^2 + (3x-5)^2 + 2x - 4(3x-5) - 5 \ldots$	5
			Method 1 & 2	**Method 1**	
			•2 express in standard quadratic form	•2 $10x^2 - 40x + 40$ } $= 0$	
			•3 factorise or use discriminant	•3 $10(x-2)^2$ } $= 0$	
			•4 interpret result to demonstrate tangency	•4 only one solution implies tangency (or repeated factor implies tangency)	
			•5 find coordinates	•5 $x = 2,\ y = 1$	
				Method 2	
				•2 $10x^2 - 40x + 40 = 0$ **stated explicitly**	
				•3 $(-40)^2 - 4 \times 10 \times 40$ or $(-4)^2 - 4 \times 1 \times 4$	
				•4 $b^2 - 4ac = 0$ so line is a tangent	
				•5 $x = 2,\ y = 1$	
			Method 3	**Method 3**	
			•1 make inference and state m_{rad}	•1 If $y = 3x - 5$ is a tangent, $m_{rad} = \frac{-1}{3}$	
			•2 find the centre and the equation of the radius	•2 $(-1, 2)$ and $3y = -x + 5$	
			•3 solve simultaneous equations	•3 $3y = -x + 5$, $y = 3x - 5$ $\rightarrow (2, 1)$	
			•4 verify location of point of intersection	•4 check (2, 1) lies on the circle.	
			•5 communicates result	•5 $\therefore$ the line is a tangent to the circle	

Question			Generic Scheme	Illustrative Scheme	Max mark
9.	(a)		•[1] know to and differentiate one term •[2] complete differentiation and equate to zero •[3] factorise derivative •[4] process for x	•[1] e.g. $f'(x) = 3x^2 ...$ •[2] $3x^2 + 6x - 24 = 0$ •[3] $3(x + 4)(x - 2)$ •[4] -4 and 2	**4**
	(b)		•[5] know how to identify where curve is increasing •[6] state range	•[5] **Method 1** 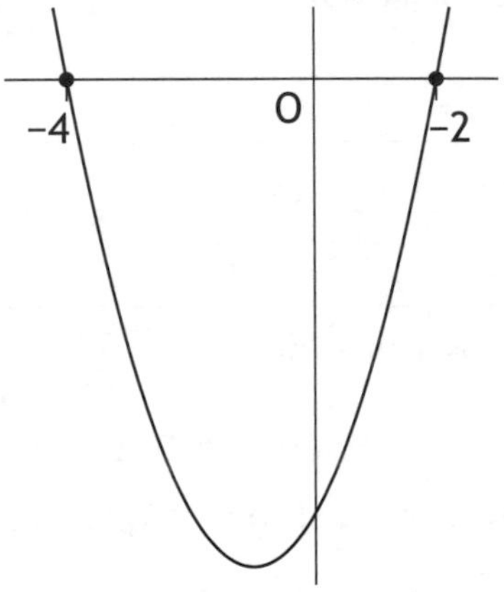 **Method 2** $3x^2 + 6x - 24 > 0$ **Method 3** Table of signs for a derivative – see page 114 for acceptable responses **Method 4** 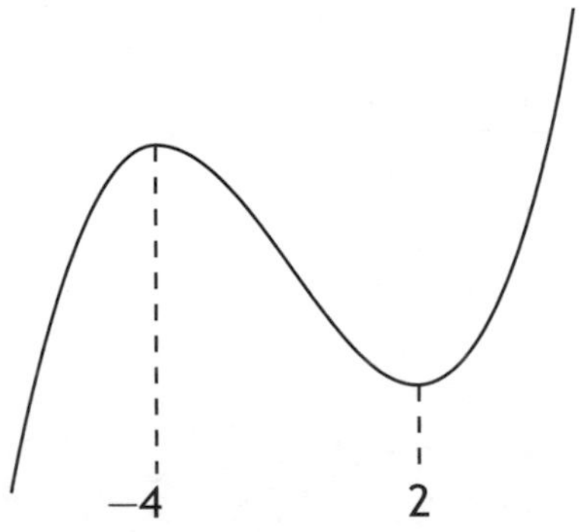 •[6] $x < -4$ and $x > 2$	**2**

Question			Generic Scheme	Illustrative Scheme	Max mark
10.			•[1] graph reflected in $y = x$ •[2] correct annotation	•[1] graph through (0, 1) and (1, 4) •[2] (0, 1) and (1, 4)	**2**
11.	(a)		•[1] interpret ratio •[2] determine coordinates	•[3] $\frac{1}{3}$ •[2] (2, 1, 0)	**2**
	(b)		•[3] find $\overrightarrow{AC}$ •[4] find $\lvert\overrightarrow{AC}\rvert$ •[5] determine k	•[3] $\overrightarrow{AC} = \begin{pmatrix} 3 \\ -6 \\ 6 \end{pmatrix}$ •[4] 9 •[5] $\frac{1}{9}$	**3**
12.	(a)		•[1] interpret notation •[2] demonstrate result	•[1] $2(3-x)^2 - 4(3-x) + 5$ •[2] $18 - 12x + 2x^2 - 12 + 4x + 5$ leading to $2x^2 - 8x + 11$	**2**
	(b)		•[3] identify common factor •[4] start to complete the square •[5] write in required form	**Method 1** •[3] $2[x^2 - 4x \ldots$ stated or implied by •[2] •[4] $2(x-2)^2 - 4 \ldots$ •[5] $2(x-2)^2 + 3$	**3**
			•[3] expand completed square form •[4] equate coefficients •[5] process for q and r and write in required form	**Method 2** •[3] $px^2 + 2pqx + pq^2 + r$ •[4] $p = 2$, $2pq = -8$, $pq^2 + r = 11$ •[5] $2(x-2)^2 + 3$	**3**

Question			Generic Scheme	Illustrative Scheme	Max mark
13.			•¹ calculate lengths AC and AD	•¹ $AC = \sqrt{17}$ and $AD = 5$ stated or implied by •³	5
			•² select appropriate formula and express in terms of p and q	•² $\cos q \cos p + \sin q \sin p$ stated or implied by •⁴	
			•³ calculate two of $\cos p$, $\cos q$, $\sin p$, $\sin q$	•³ $\cos p = \frac{4}{\sqrt{17}}$, $\cos q = \frac{4}{5}$ $\sin p = \frac{1}{\sqrt{17}}$, $\sin q = \frac{3}{5}$	
			•⁴ calculate other two and substitute into formula	•⁴ $\frac{4}{5} \times \frac{4}{\sqrt{17}} + \frac{3}{5} \times \frac{1}{\sqrt{17}}$	
			•⁵ arrange into required form	•⁵ $\frac{19}{5\sqrt{17}} \times \frac{\sqrt{17}}{\sqrt{17}} = \frac{19\sqrt{17}}{85}$ or $\frac{19}{5\sqrt{17}} = \frac{19\sqrt{17}}{5 \times 17} = \frac{19\sqrt{17}}{85}$	
14.	(a)		•¹ state value	•¹ 2	1
	(b)		•² use result of part (a)	•² $\log_4 x + \log_4 (x - 6) = 2$	5
			•³ use laws of logarithms	•³ $\log_4 x(x - 6) = 2$	
			•⁴ use laws of logarithms	•⁴ $x(x - 6) = 4^2$	
			•⁵ write in standard quadratic form	•⁵ $x^2 - 6x - 16 = 0$	
			•⁶ solve for x and identify appropriate solution	•⁶ 8	
15.	(a)		•¹ state value of a	•¹ $a = 4$	3
			•² state value of b	•² $b = -5$	
			•³ calculate k	•³ $k = -\frac{1}{12}$	
	(b)		•⁴ state range of values	•⁴ $d > 9$	1

Paper 2

Question			Generic Scheme	Illustrative Scheme	Max mark
1.	(a)	(i)	•[1] state the midpoint M	•[1] (2, 4)	1
		(ii)	•[2] calculate gradient of median •[3] determine equation of median	•[2] 4 •[3] $y = 4x - 4$	2
	(b)		•[4] calculate gradient of PR •[5] use property of perpendicular lines •[6] determine equation of line	•[4] 1 •[5] -1 stated or implied by •[6] •[6] $y = -x + 6$	3
	(c)			**Method 1**	3
			•[7] find the midpoint of PR •[8] substitute $x(y)$-coordinate into equation of L •[9] verify $y(x)$-coordinate and communicate conclusion	•[7] (5, 1) •[8] $y = -5 + 6$ $(1 = -x + 6)$ •[9] $y = 1 (x = 5)$ $\therefore$ L passes through the midpoint of PR	
				Method 2	3
			•[7] find the midpoint of PR •[8] substitute x and y coordinates into the equation of L •[9] verify the point satisfies the equation and communicate conclusion	•[7] $x + y = 6$ sub (5,1) •[8] $5 + 1 = 6$ •[9] $\therefore$ point (5, 1) satisfies equation	
				Method 3	3
			•[7] find the midpoint of PR •[8] find equation of PR •[9] use simultaneous equations and communicate conclusion	•[7] (5, 1) •[8] $y = x - 4$ •[9] $y = 1$, $x = 5$ $\therefore$ L passes through the midpoint of PR	
				Method 4	3
			•[7] find the midpoint of PR •[8] find equation of perpendicular bisector of PR •[9] communicate conclusion	•[7] (5, 1) •[8] $y - 1 = -1(x - 5) \rightarrow y = -x + 6$ •[9] The equation of the perpendicular bisector is the same as L therefore L passes through the midpoint of PR	

Question			Generic Scheme	Illustrative Scheme	Max mark
2.			•[1] use the discriminant •[2] simplify and apply the condition for no real roots •[3] state range	•[1] $(-2)^2 - 4(1)(3 - p)$ •[2] $-8 + 4p < 0$ •[3] $p < 2$	3
3.	(a)	(i)	 •[1] know to substitute $x = -1$ •[2] complete evaluation, interpret result and state conclusion	**Method 1** •[1] $2(-1)^3 - 9 \times (-1)^2 + 3 \times (-1) + 14$ •[2] $= 0 \therefore (x + 1)$ is a factor	2
			 •[1] know to use $x = -1$ in synthetic division •[2] complete division, interpret result and state conclusion	**Method 2** •[1] $\begin{array}{r\|rrrr} -1 & 2 & -9 & 3 & 14 \\ & & -2 & & \\ \hline & 2 & -11 & & \end{array}$ •[2] $\begin{array}{r\|rrrr} -1 & 2 & -9 & 3 & 14 \\ & & -2 & 11 & -14 \\ \hline & 2 & -11 & 14 & 0 \end{array}$ remainder $= 0 \therefore (x + 1)$ is a factor	2
			 •[1] start long division and find leading term in quotient •[2] complete division, interpret result and state conclusion	**Method 3** •[1] $2x^2$ $(x+1)\overline{)2x^3 - 9x^2 + 3x + 14}$ •[2] $2x^2 - 11x + 14$ $(x+1)\overline{)2x^3 - 9x^2 + 3x + 14}$ $\underline{2x^3 + 2x^2}$ $-11x^2 + 3x$ $\underline{-11x^2 - 11x}$ $14x + 14$ $\underline{14x + 14}$ 0 remainder $= 0 \therefore (x + 1)$ is a factor	2
		(ii)	•[3] state quadratic factor •[4] find remaining linear factors or substitute into quadratic formula •[5] state solution	•[3] $2x^2 - 11x + 14$ •[4] $\ldots(2x - 7)(x - 2)$ **or** $\dfrac{11 \pm \sqrt{(-11)^2 - 4 \times 2 \times 14}}{2 \times 2}$ •[5] $x = -1,\ 2,\ 3{\cdot}5$	3

Question			Generic Scheme	Illustrative Scheme	Max mark
	(b)	(i)	•6 state coordinates	•6 (−1,0) and (2,0)	1
		(ii)	•7 know to integrate with respect to x •8 integrate •9 interpret limits and substitute •10 evaluate integral	•7 $\int\left(2x^3-9x^2+3x+14\right)dx$ •8 $\frac{2x^4}{4}-\frac{9x^3}{3}+\frac{3x^2}{2}+14x$ •9 $\left(\frac{2\times2^4}{4}-\frac{9\times2^3}{3}+\frac{3\times2^2}{2}+14\times2\right)-\left(\frac{2\times(-1)^4}{4}-\frac{9\times(-1)^3}{3}+\frac{3\times(-1)^2}{2}+14\times(-1)\right)$ •10 27	4
4.	(a)		•1 centre of C_1 •2 radius of C_1 •3 centre of C_2 •4 radius of C_2	•1 (−5, 6) •2 3 •3 (3, 0) •4 5	4
	(b)		•5 calculate the distance between the centres •6 calculate the sum of the radii •7 interpret significance of calculations	•5 10 •6 8 •7 $8<10$ ∴ the circles do not intersect	3
5.	(a)		•1 find $\overrightarrow{AB}$ •2 find $\overrightarrow{AC}$	•1 $\begin{pmatrix}-8\\16\\2\end{pmatrix}$ •2 $\begin{pmatrix}-2\\-8\\16\end{pmatrix}$	2
	(b)		**Method 1** •3 evaluate $\overrightarrow{AB}.\overrightarrow{AC}$ •4 evaluate $\lvert\overrightarrow{AB}\rvert$ and $\lvert\overrightarrow{AC}\rvert$ •5 use scalar product •6 calculate angle	**Method 1** •3 $\overrightarrow{AB}.\overrightarrow{AC}=16-128+32=-80$ •4 $\lvert\overrightarrow{AB}\rvert=\lvert\overrightarrow{AC}\rvert=18$ •5 $\cos BAC=\frac{-80}{18\times18}$ •6 $104\cdot3^\circ$ or 1·82 radians	4
			Method 2 •3 calculate length of BC •4 calculate lengths of AB and AC •5 use cosine rule •6 calculate angle	**Method 2** •3 $BC=\sqrt{808}$ •4 $AB=AC=18$ •5 $\cos BAC=\frac{18^2+18^2-\sqrt{808}^2}{2\times18\times18}$ •6 $104\cdot3^\circ$ or 1·82 radians	4

Question			Generic Scheme	Illustrative Scheme	Max mark
6.	(a)		•1 state the number	•1 200	1
	(b)		•2 interpret context and form equation •3 know to and use logarithms appropriately •4 simplify •5 evaluate t	•2 $2 = e^{0\cdot107t}$ •3 $\ln 2 = \ln (e^{0\cdot107t})$ •4 $\ln 2 = 0\cdot107t$ •5 $t = 6\cdot478 \ldots$	4
7.	(a)		•1 expression for length in terms of x and y •2 obtain an expression for y •3 demonstrate result	•1 $9x + 8y$ •2 $y = \frac{108}{6x}$ •3 $L(x) = 9x + 8\left(\frac{108}{6x}\right)$ leading to $L(x) = 9x + \frac{144}{x}$	3
	(b)		•4 know to and start to differentiate •5 complete differentiation •6 set derivative equal to 0 •7 solve for x •8 verify nature of stationary point •9 interpret and communicate result	•4 $L'(x) = 9\ldots$ •5 $L'(x) = 9 - \frac{144}{x^2}$ •6 $9 - \frac{144}{x^2} = 0$ •7 $x = 4$ •8 Table of signs for a derivative – see page 114 for acceptable responses •9 Minimum at $x = 4$ **or** •8 $L''(x) = \frac{288}{x^3}$ •9 $L''(4) > 0 \therefore$ minimum Do not accept $\frac{d^2y}{dx^2} = \ldots$	6
8.	(a)		•1 use compound angle formula •2 compare coefficients •3 process for k •4 process for a and express in required form	•1 $k \cos x \cos a - k \sin x \sin a$ **stated explicitly** •2 $k \cos a = 5$, $k \sin a = 2$ **stated explicitly** •3 $k = \sqrt{29}$ •4 $\sqrt{29} \cos(x + 0\cdot38)$	4

Question			Generic Scheme	Illustrative Scheme	Max mark
	(b)		•5 equate to 12 and simplify constant terms •6 use result of part (a) and rearrange •7 solve for $x + a$ •8 solve for x	•5 $5\cos x - 2\sin x = 2$ or $5\cos x - 2\sin x - 2 = 0$ •6 $\cos(x + 0{\cdot}3805\ \ldots) = \frac{2}{\sqrt{29}}$ •7 $1{\cdot}1902\ \ldots$, •8 $5{\cdot}0928\ \ldots$ •7 $1{\cdot}1902\ \ldots$, $5{\cdot}0928\ \ldots$ •8 $0{\cdot}8097\ \ldots$, $4{\cdot}712\ \ldots$	4
9.			•1 write in integrable form •2 integrate one term •3 complete integration •4 state expression for $f(x)$	•1 $2x^{\frac{1}{2}} + x^{-\frac{1}{2}}$ •2 $\frac{4}{3}x^{\frac{3}{2}}$ or $2x^{\frac{1}{2}}$ •3 $2x^{\frac{1}{2}} + c$ or $\frac{4}{3}x^{\frac{3}{2}} + c$ •4 $f(x) = \frac{4}{3}x^{\frac{3}{2}} + 2x^{\frac{1}{2}} - 2$	4
10.	(a)		•1 start to differentiate •2 complete differentiation	•1 $\frac{1}{2}(x^2+7)^{-\frac{1}{2}}\ldots$ •2 $\ldots \times 2x$	2
	(b)		•3 link to (a) and integrate	•3 $4(x^2+7)^{\frac{1}{2}}\ (+\ c)$	1
11.	(a)		•1 substitute for sin $2x$ and tan x •2 simplify •3 use an appropriate substitution •4 simplify and communicate result	•1 $(2\sin x\cos x) \times \frac{\sin x}{\cos x}$ •2 $2\sin^2 x$ •3 $2(1 - \cos^2 x)$ **or** $1 - (1 - 2\sin^2 x)$ •4 $1 - \cos 2x = 1 - \cos 2x$ **or** $2\sin^2 x = 2\sin^2 x$ $\therefore$ identity shown	4
	(b)		•5 link to (a) and substitute •6 differentiate	•5 $f(x) = 1 - \cos 2x$ **or** $f(x) = 2\sin^2 x$ •6 $f'(x) = 2\sin 2x$ **or** $f'(x) = 4\sin x\cos x$	2

Table of signs for a derivative – acceptable responses

x	-4^-	-4	-4^+
$\frac{dy}{dx}$ or $f'(x)$	+	0	–
Shape or slope	/	—	\

x	2^-	2	2^+
$\frac{dy}{dx}$ or $f'(x)$	–	0	+
Shape or slope	\	—	/

x	$\rightarrow$	-4	$\rightarrow$
$\frac{dy}{dx}$ or $f'(x)$	+	0	–
Shape or slope	/	—	\

x	$\rightarrow$	2	$\rightarrow$
$\frac{dy}{dx}$ or $f'(x)$	–	0	+
Shape or slope	\	—	/

Arrows are taken to mean “in the neighbourhood of”.

x	a	-4	b	2	c
$\frac{dy}{dx}$ or $f'(x)$	+	0	–	0	+
Shape or slope	/	—	\	—	/

Where: $a < -4$, $-4 < b < 2$, $c > 2$.

Since the function is continuous “$-4 < b < 2$” is acceptable.

x	$\rightarrow$	-4	$\rightarrow$	2	$\rightarrow$
$\frac{dy}{dx}$ or $f'(x)$	+	0	–	0	+
Shape or slope	/	—	\	—	/

Since the function is continuous “$-4 \rightarrow 2$” is acceptable.

Acknowledgements

Hodder Gibson would like to thank the SQA for use of any past exam questions that may have been used in model papers, whether amended or in original form.